Electronic Devices and Circuit Applications

Roger W. Prewitt
Stephen W. Fardo

Reston Publishing Company, Inc.
A Prentice-Hall Company
Reston, Virginia

Library of Congress Cataloging in Publication Data

Prewitt, Roger W.
    Electronic devices and circuit applications.

    1. Electronics. I. Fardo, Stephen W. II. Title.
TK7816.P68   1984   621.381   84-15961
ISBN 0-8359-1645-6

To our parents

Ernest and Ada Prewitt
Hollis and Velma Fardo

for their kindness and understanding throughout our lives

© 1985 by Reston Publishing Company, Inc.
A Prentice-Hall Company
Reston, Virginia 22090

10 9 8 7 6 5 4 3 2 1

PRINTED IN THE UNITED STATES OF AMERICA

Contents

Preface

Electronic Devices and Circuit Applications is a comprehensive text designed for use in a one-semester or two-trimester course sequence in an electronics program. College, university, and vocational-technical training programs should find this introductory text to be most informative and easy to understand. A practical, simplified approach is followed in teaching the basic concepts used in all areas of electronics. The mathematical content is presented by using basic operations, with no requirement for advanced mathematics.

The book is organized into 16 chapters, with many types of devices and circuit applications discussed. The book is organized so that key concepts build on one another. Introductions to digital circuits and microprocessors are included as applications of integrated circuits, and there is a separate chapter dealing with electronic device testing. The content is up-to-date and comprehensive.

To enhance student learning, many practical examples are used throughout the book. The authors have had over 25 years of combined teaching experience at the university level and have been actively involved in teaching electronic devices and circuit applications in a large electronics technology program. Their experience has helped them to develop many practical examples and problems which aid student understanding. The authors' experience in textbook writing includes a combined total of 14 publications dealing with various areas of electricity and electronics.

The way in which this text is used in electronics technology programs will vary, depending on specific course organization within a technical program. There are many chapter combinations which could be devised. The most important aspect of this book is its practical coverage of electronic devices and circuit applications. The basic approach, practical examples, and abundance of problems make the text very desirable in electronics technology programs. Most chapters are organized as follows: introduction, major content, review, problems, and suggested lab activities. This organization allows the student to progress through a chapter systematically. The introduction provides an overview of the chapter content and objectives. The major content of each chapter is written in a manner that is easy to understand. The review section consists of questions and discussion items that emphasize the major content areas of the chapter. The problems at the end of each chapter are used to supplement the understanding of electronic devices and circuit applications through problem solving.

Roger W. Prewitt
Stephen W. Fardo

CHAPTER 1
Electronic Devices

Understanding the movement of electrons within electronic devices is very important to the student who attempts to gain knowledge concerning how and why electronic devices function. In this chapter current flow in semiconductor materials, vacuums, and gases is examined.

CONDUCTORS, INSULATORS, AND SEMICONDUCTORS

A material through which current flows easily is called a *conductor*. Copper and aluminum wire are commonly used as conductors. Conductors have low resistance to electric current flow. Several metals are considered to be electric conductors. Each type of metal has a different ability to conduct electric current. For example, silver is a better conductor than copper, but silver is too expensive to use in large quantities. Aluminum does not conduct electric current as well as copper, but it is cheaper and more lightweight. Copper is the most commonly used conductor material.

Some materials do not allow electric current to flow easily through them. *Insulators* have high resistance to the movement of electric current. Some examples of insulators are plastic and rubber.

Materials called *semiconductors* are used extensively in the manufacture of electronic devices. Semiconductor materials are not conductors and not insulators. Some common types of semiconductor materials are silicon, germanium, and selenium. A comparison of conductors, insulators, and semiconductors is shown in Figure 1–1.

Semiconductor material is a material that is classified electrically on a scale between those materials that are said to be insulators and those classified as conductors. Semiconductor material is not classified

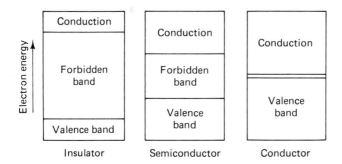

FIGURE 1–1   The valence band of an atom contains electrons which may move through the "forbidden band" and go into conduction with sufficient energy.

as a good conductor of electric current, nor is it classified as a good insulator to the flow of electric current.

CURRENT FLOW IN SEMICONDUCTOR MATERIAL

To understand how current flows in semiconductor materials, one must examine the atomic bonding and structure of the material itself. All materials exist due to a very specialized arrangement of their atoms. This arrangement, known as *atomic bonding,* may be classified as ionic, metallic, or covalent bonding and deals with how atoms join other atoms to form molecules of materials. Covalent bonding is most important in the study of semiconductor principles.

The outer layer, or shell, of all atoms contains from one to eight electrons. These electrons are known as valence electrons and their number determines whether an atom is stable or unstable. Atoms with eight valence electrons are classified as stable atoms and generally will not join with other atoms to form molecules. Atoms with fewer than eight valence electrons are classified as unstable atoms and will readily join with other atoms to establish stability.

The atomic bonding that allows atoms to share valence electrons for stability is known as *covalent bonding.* The two atoms illustrated in Figure 1–2 represent a covalent bond. It should be noted that each atom in this illustration has four valence electrons. When these two atoms share their valence electrons, the total number of shared electrons in their combined outer layers becomes eight and stability is achieved.

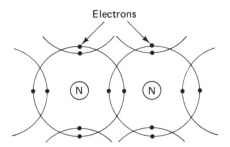

FIGURE 1–2　Covalent bonding of two atoms.

Silicon and germanium are the most frequently used semiconductor materials in the electronics industry. An atom of either material has four valence electrons. The atoms of both materials form covalent bonds for stability. Both silicon and germanium are very good insulators when in their purest form.

In order for silicon or germanium to become classified as a semiconductor material rather than as an insulator, the content of its covalent bond must be altered. This is accomplished in a carefully controlled fashion by adding foreign atoms, known as *impurities,* to the existing covalent structure of silicon or germanium. This addition of impurities is known as *doping.* For example, let us examine the change in the structure of the covalent bond when an impurity such as an arsenic atom covalently bonds with a germanium atom. The arsenic atom has five valence electrons, while the germanium atom has four valence electrons. The four valence electrons from the germanium atom will covalently bond with four of the five valence electrons from the arsenic atom. This results in the fifth electron from the arsenic atom being omitted from any covalent bond, thus becoming an "extra," or *"free," electron* (free from a covalent bond). If many impurities that have five valence electrons are added to many germanium atoms, the result is a semiconductor material with many, many free electrons. This type of semiconductor material is classified as N *material.* The impurity that added electrons to the overall bonding structure is called a *donor* because it donated electrons, causing free electrons to exist, as illustrated in Figure 1–3.

Likewise let us examine the change in the covalent bond when an impurity atom with three valence electrons (gallium or indium) is bonded with an atom of germanium. In this case three of the four valence electrons from the germanium atom would bond with the three valence electrons from the impurity. This would result in a space where an electron should be but is not. This space is called a *"hole"* and is illustrated in Figure 1–4.

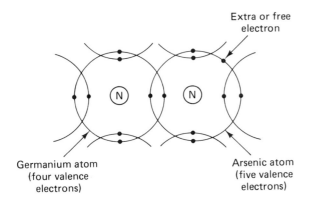

FIGURE 1–3   Free electrons resulting from donor.

If many impurities that have three valence electrons are added to many germanium atoms, the result is a semiconductor material with many many spaces where electrons should be located but are not. This type of semiconductor material is classified as *P material*. The impurity added to the bonding structure that causes an overall deficiency in electrons or an excessive number of holes is called an *acceptor*. Thus, when a donor impurity is doped with germanium, an N material results with an excessive number of electrons. When an acceptor impurity is doped with germanium, the result is a material deficient in electrons.

Both the N- and P-type semiconductor materials have current carriers. Current carriers are responsible for current flow within the semiconductor materials. These current carriers are classified as *majority carriers* and *minority carriers*. Both majority and minority carriers exist in both types of semiconductor materials.

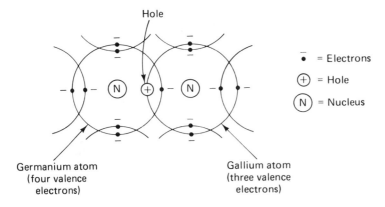

FIGURE 1–4   Holes resulting from acceptor.

In an N material there are many free electrons due to doping. Because of imperfections in the doping process, there are also a few holes. Since there are many free electrons and very few holes, electrons are classified as majority current carriers and holes are classified as minority current carriers in all N-type semiconductor material.

The opposite is true with P-type semiconductor materials. The P material has many holes (spaces where electrons should be but are not) due to the doping process. Likewise because of imperfections in the doping process, all P material has a few free electrons. Since there are many holes and few free electrons, holes are classified as majority current carriers and free electrons are classified as minority current carriers in all P-type semiconductor material.

Figure 1–5 illustrates the effect upon the majority and minority current carriers of an N-type semiconductor material when influenced by an external voltage source.

The external current (*I*) flowing to and from the voltage source is mainly due to the action of the majority carriers. Electrons are negative (−) charges. A deficiency in electrons is said to be a positive (+) charge. Thus, the free electrons are negative charges and are attracted to the positively charged terminal of the voltage source, while the holes are positive charges and are attracted to the negatively charged terminal of the voltage source. As the electrons are drawn toward the positive voltage source, new holes are created in the space once occupied by these electrons. As electrons move from the N material (as external current *I* toward the positive voltage terminal), electrons move onto the N material from the negative voltage terminal to occupy the holes. There is very little external current due to the action of the minority carriers.

Figure 1–6 illustrates the effect upon the majority and minority current carriers of a P-type material when influenced by a similar voltage.

Again the external current flowing to and from the voltage source

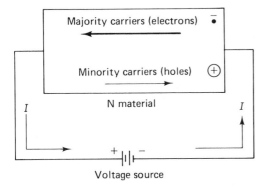

FIGURE 1–5 Majority-minority carriers in N material.

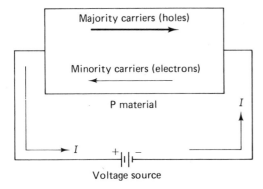

FIGURE 1–6   Majority-minority carriers in P material.

is largely due to the action of the majority current carriers, which in this case are holes. Since holes are where electrons should be located but are not, they are considered to be positively charged and attracted toward the negative terminal of the voltage source. As electrons move onto the P material from the negative voltage terminal to fill the holes, electrons will move from the P material due to the attractive force of the positive voltage terminal. As before, there is very little external current due to the action of the minority carriers. When P and N materials are formed on the same piece of silicon or germanium in various combinations, electronic devices such as *diodes* and *transistors* are produced.

CURRENT FLOW IN VACUUMS AND GASES

Current flow in a vacuum deals with the movement of electrons within a chamber that is void of normal oxygen or air. In addition to voltage, current flow in a vacuum is also affected and controlled by heat energy. Current flow in a vacuum may also be controlled by light energy.

Current flow in a gas refers to the movement of electrons within an enclosure that has had all of the oxygen or air replaced by an inert gas such as argon. In addition to voltage, current flow can be affected and controlled by heat energy, as well as by the ionization of the gas. As in a vacuum, current flow in a gas can also be controlled by light energy.

CURRENT FLOW IN A VACUUM

Thomas A. Edison discovered that current would flow through space between heated and charged metal conductors when placed in a closed chamber. He made this discovery while attempting to perfect his electric light. This unexplained action of current was called the Edison effect. Later Sir J. J. Thompson explained the Edison effect as electrons emitted from a heated surface and attracted through space to a positively charged conductor. What these gentlemen discovered led to the development of the vacuum tube.

One means by which current moves through a vacuum is caused by thermionic emission. *Thermionic emission* results when certain metals are heated and electrons are emitted into the space immediately around the metal's surface. Generally, the electrons associated with atoms of metals are held so tightly to the parent atom that they will not leave the metal's surface. However, when the atoms found in certain metals are heated, their electrons gain enough energy from the heat to actually escape and fly off into space. Figure 1–7 illustrates thermionic emission.

In this illustration a *filament* of wire is enclosed in a glass bulb. The oxygen has been removed from the bulb to prevent the filament from being destroyed when heated. This is due to the rapid oxidation of the metal that normally would take place. Even though the filament is completely enclosed inside the bulb, external connections through the glass permit a voltage source to be attached. When attached to the filament, the voltage source causes a current to flow through the filament. This current flow causes the filament to heat up much like the filament of an incandescent light bulb, although not as bright. This heat provides the electrons enough additional energy to cause them to be emitted into the space around the filament. These electrons form a "space cloud" around the heated filament. When the number of electrons forming the space cloud is large enough, the space cloud takes on a negative charge strong enough to prevent additional electrons from leaving the filament's surface; thus, electron emission ceases. This is sometimes referred to as *space cloud saturation* and is determined by the type of

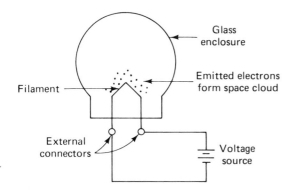

FIGURE 1–7   Electrons emitted due to thermionic emission.

metal used in the construction of the filament, the amount of current flow through the filament, the amount of heat caused by this current, and the physical size of the filament. Once space cloud saturation has been reached, the amount of current flow through the filament may be increased to raise the temperature (within limits) of the filament. This results in additional electrons being emitted, an expansion of the space cloud, and a new space cloud saturation point being reached.

Keeping in mind that electrons are negative by nature and that unlike charges attract each other, a positively charged conductor may be placed near the space cloud, within the glass enclosure, to attract these electrons. Figure 1–8 illustrates how this might be accomplished.

In this illustration an additional conductor has been inserted and sealed within the glass enclosure. An external connection has been provided that allows connection of voltage source 2 between the filament and the additional conductor, which is called the *plate*, or collector. Voltage source 2 is connected in such a way as to cause the plate to be positive. Voltage source 1 provides the means to allow current to flow through the filament to provide heat as in the past. As the filament heats, electrons are emitted into the vacuum. The strong positive charge on the plate attracts the negatively charged electrons, thus causing them to flow from the enclosure as external current. Current will flow from the heated filament through the vacuum gap, between the filament and plate, onto the plate, and to the positive terminal of voltage source 2 as long as conditions remain unchanged. If the plate becomes more positive, more electrons will be attracted to it and the external current flow to the positive terminal of voltage source 2 will increase accordingly. This will be true only until the plate becomes so positive that every electron emitted

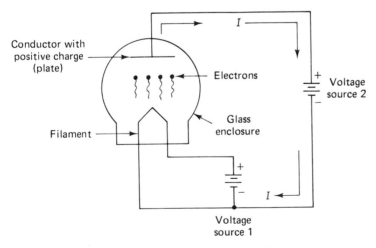

FIGURE 1–8 Electron attraction to positive charge.

by the filament is attracted to the plate. At this point plate saturation is realized. Plate saturation is a condition that prevails when increasing the positive charge on the plate no longer brings about an increase in the current flow through the glass enclosure.

Photoelectric Emission

Another cause of current flow through a vacuum can be linked to *photoelectric emission*. Some materials will actually emit electrons when exposed to an appropriate light source. If this light-sensitive material is placed within a sealed glass enclosure with all of the oxygen removed and is exposed to an appropriate light source, electrons will be emitted into the space around the light-sensitive material.

Figure 1–9 illustrates how this might be possible. In this illustration a metallic surface called a *cathode* has been coated with a light-sensitive material and is sealed within the vacuum glass enclosure. An external connection is provided for the coated cathode. When an appropriate light is caused to fall upon or strike the cathode's surface, electrons are emitted into space. Since light is a form of energy, it provides the additional energy needed by the electrons to leave the coated surface of the cathode. As the intensity of the light falling on the cathode is increased, the number of electrons being emitted by the cathode will increase. This is true until the coated cathode reaches the point of space cloud saturation previously described.

If a positively charged plate is inserted into the glass enclosure, the electrons that are emitted by the coated cathode will be attracted to the plate and will appear as external current. This is illustrated in Figure 1–10. In this illustration the positively charged plate is positioned near the coated cathode. An appropriate light is caused to fall on the light-sensitive coating of the cathode. This causes the cathode to emit electrons, which are attracted to the positive plate located within the glass enclosure. These electrons appear as external current flowing to and from the voltage source. This current will continue to flow through the

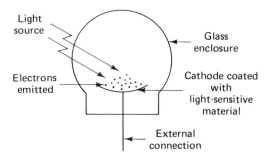

FIGURE 1–9   Electrons emitted due to light.

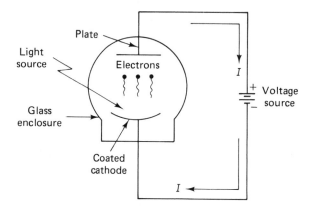

FIGURE 1–10   Electrons attracted by positive charge.

vacuum gap between the cathode and plate as long as existing conditions remain unchanged. If the light intensity is increased, an increase in the current flow between the cathode and plate will occur. An increased voltage between the cathode and plate will cause immediate plate saturation for any amount of light intensity.

CURRENT FLOW IN A GAS

Current flow in a gas is somewhat similar in action to current flow in a vacuum. The difference lies in the characteristics of the gas and the reaction of the gas atoms when struck by moving electrons.

In Figure 1–11 a sealed chamber has all of the air within it carefully replaced by a gas such as neon, argon, or mercury vapor. Into this

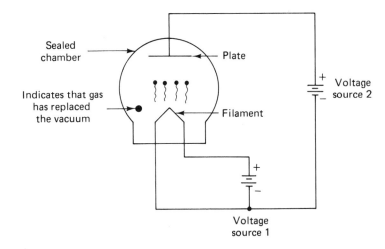

FIGURE 1–11   Electrons emitted in a gas.

chamber are inserted a filament and a plate. The filament is heated by voltage source 1 and emits electrons. The plate is positive because of voltage source 2 and attracts these emitted electrons. As the electrons move toward the plate with sufficient velocity, they strike the atoms of the gas placed in the chamber. This striking action causes additional electrons to be knocked from the atoms. The electrons that are knocked from the atoms are attracted to the positive plate and join the electrons that are emitted by the filament, as all electrons travel in that direction.

It should be noted at this point that when, for any reason, an atom acquires more electrons than it should have, it is classified as a *negative ion*. Likewise if an atom loses an electron, it is classified as a *positive ion*. Thus, when many electrons emitted by the filament in our example strike many gas atoms causing them to lose electrons, many positive ions are created. When this happens the gas is said to *ionize*. The positive gas ions created because of ionization are attracted to the negative space cloud near the filament. These positive gas ions generally regain their lost electrons as they travel toward the filament or at the space cloud itself.

The voltage between the plate and filament, illustrated in our example, must be great enough to cause the gas to ionize. This voltage is called the "turn-on," ignition, or *ionization voltage*. After the ignition voltage is provided and ionization takes place, a smaller voltage between the plate and filament will maintain ionization. If the voltage between the plate and filament drops below a certain critical level, the gas will deionize and current will no longer flow.

Just as light energy is a cause of current flow in a vacuum, likewise light can cause current flow in a gas. If a light-sensitive cathode was placed in a sealed chamber of gas, rather than a vacuum chamber, ionization and current flow would take place as previously described.

Figure 1–12 illustrates ionization of gas due to electrons being

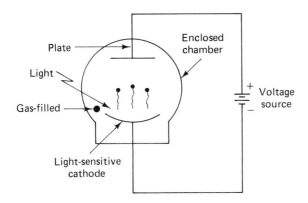

FIGURE 1–12   Ionization due to light.

emitted because of light. When light strikes the light-sensitive cathode, electrons are emitted just as they were in a vacuum. These electrons are negative and are attracted toward the positive plate. As they travel toward the plate, the electrons that have been emitted because of light strike the gas atoms and cause ionization. The electrons that are knocked from the gas atoms due to ionization join the emitted electrons as all travel toward the plate. The positive ions are attracted toward the cathode and regain lost electrons as they travel to the cathode or at the cathode itself. For current to flow in a gas due to light, the light must be intense enough to cause the cathode to emit a sufficient number of electrons. Likewise, the voltage between the plate and cathode must be large enough to cause these emitted electrons to travel at a sufficient velocity to bring about ionization.

Current flow in a gas can be caused by voltage only as illustrated in Figure 1–13. A special cathode, known as a *"cold cathode"* because it is not heated and will not emit electrons due to light, is placed in a sealed enclosure with gas. A positive plate is placed very near this cathode in the same enclosure. When the voltage between this plate and cathode becomes great enough, the gas will ionize, causing current to flow through the gas between the cathode and plate.

The atoms of a gas are constantly moving about in the chamber. Because of this movement, gas atoms will bump into each other, causing a few electrons to be knocked from these atoms. This causes the existence of a few naturally occurring "free" electrons in the gas. A high voltage between the cold cathode and the plate will attract these free electrons toward the positive plate. As these electrons move toward the plate, they crash into gas atoms, causing additional electrons to be knocked from these atoms. These additional electrons join the others as all travel toward the plate. The positive ions that are created are attracted to the cathode. As these positive ions strike the cathode, addi-

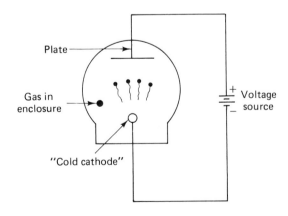

FIGURE 1–13   Cold-cathode emission.

tional electrons are knocked from the cathode and are attracted toward the plate. This entire action increases until the gas ionizes or ignites. When current flow in a gas is caused in this manner, a relatively high voltage is required. All electron gaseous tubes exhibit the characteristics of current flow in a gas as just described.

In this chapter current flow in semiconductor materials, vacuums, and gases was examined. The performance of all electronic devices relies upon one of these types of electron movement.

REVIEW

1. What is meant by covalent bonding in reference to semiconductor materials?

2. What is the relation between doping and semiconductor materials?

3. What is the difference between a donor impurity and an acceptor impurity?

4. Why are some semiconductor materials classified as P type materials while others are classified as N type?

5. What are majority and minority carriers of current?

6. What are "holes" as associated with semiconductor materials?

7. What is the relation between semiconductor material and electronic devices?

8. What contribution did Thomas A. Edison make toward the development of the electron vacuum tube?

9. What is thermionic emission and what is its relation to current flow in a vacuum?

10. Why is a space cloud described as being negative?

11. How can light cause current flow in a vacuum?

12. What is the difference between a positive and negative ion?

13. What is the difference between ignition and deionization?

14. What is a "cold cathode"?

15. What is the relation between current flow in a gas and the field of electronics?

16. What is the relation between current flow in a vacuum and the field of electronics?

CHAPTER 2
Diodes

Diodes are among the most commonly used devices in electronics today. Understanding how diodes perform is important in developing a knowledge of the operation and characteristics of most other electronic devices. In the following sections the characteristics of the basic PN junction semiconductor signal and power diodes, as well as the characteristics of more specialized diodes, including the zener, varicap (varactor), and tunnel diodes, are discussed. Vacuum and gaseous diodes are also discussed.

PN JUNCTION CHARACTERISTICS

In Chapter 1 current flow in semiconductor materials was investigated. Recall that there are two basic materials used in the manufacture of semiconductor electronic devices. These are P- and N-type materials. Both materials rely upon majority and minority current carriers to allow an external current flow. The majority carriers in a P material are holes, while the majority carriers in an N material are electrons. The minority carriers in a P material are electrons while the minority carriers in an N material are holes. External current through either material is the result of the action of the majority carriers more so than the minority carriers.

When one side of a chip of silicon or germanium is doped as a P material and the other side is doped as an N material, a PN junction diode is produced. Figure 2–1 illustrates this arrangement along with the electrical symbol of a PN junction diode. In this illustration it can be seen that there are two external leads connected to this device. One lead

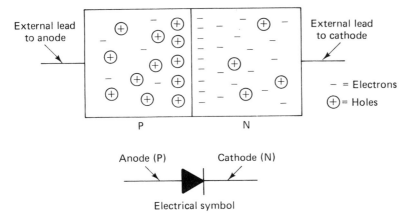

FIGURE 2–1   PN junction diode.

is affixed to the P side of the diode while the other is connected to the N side. Likewise this illustration shows that the N side of the diode is called the *cathode* while the P side is called the *anode*. Finally, as illustrated, there are many holes and few electrons on the P side, while there are many electrons and few holes on the N side.

The P and N materials are normally separated by a junction where one material ends and the other begins. Around and on either side of the junction is an area that is void of current carriers. This area is known as the depletion area, or the *depletion zone*, and is a natural formation that results from the influence of the current carriers from both the N and P materials. Figure 2–2 illustrates this depletion area.

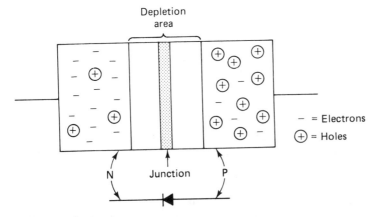

FIGURE 2–2   Depletion zone.

At the instant the N and P materials are formed, carriers around the junction react. Majority carriers from the P material move across the junction into the N material while majority carriers from the N material move across the junction into the P material. In both instances the electrons combine with holes, resulting in the depletion of majority carriers in this region. The action of the few minority carriers that are found near the junction is insignificant even though these carriers do move into opposite materials, resulting in total depletion of carriers in the area around the junction.

Since electrons (majority carriers) from the N material move to the P side of the junction, this creates an unusual positive condition on the N material near the junction. Remember, the absence of electrons, where electrons should be located, is said to be positive. Likewise the movement of holes (majority carriers) from the P material to the N side of the junction creates a negative condition on the P material near the junction. An excessive number of electrons is said to cause a negative state. This positive condition on the N material, once established, forms a positive barrier and repels any additional movement of holes onto this material, while the negative condition established on the P material forms a negative barrier and likewise repels any additional movement of electrons onto the P side of the junction. The action of the carriers around the junction results in the formation of a small difference of potential (a few tenths of a volt) between the N and P material near the junction. This difference in potential is called the *potential barrier* and is illustrated in Figure 2–3.

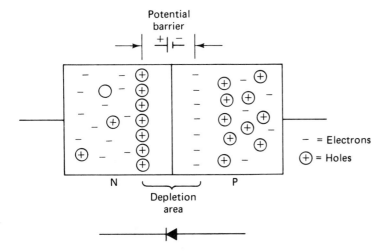

FIGURE 2–3   Potential barrier.

DIODE BIASING

Due to the arrangement of the current carriers as previously described, the PN junction diode acts like a single-pole, single-throw (SPST) switch when a voltage of appropriate magnitude and polarity is applied to its external leads. Providing a voltage to the external leads of a PN junction diode to control its action is known as *biasing*. The voltage that is used to control the action of the diode is known as *biasing voltage*.

FORWARD BIASING

A PN junction diode may be forward biased or reverse biased. When the anode (P material) of a diode is made positive and the cathode (N material) is made negative, the diode is said to be *forward biased*. When forward biased the diode exhibits a very low forward resistance and readily conducts, allowing an external forward current flow. This is true as long as the biasing voltage is larger than the few tenths of a volt associated with the potential barrier.

Figure 2–4 illustrates the PN junction diode when forward biased. The voltage provided by the battery in this illustration is large enough to overcome the potential barrier. The negative terminal of the battery is connected to the cathode (N material) of the diode while the positive terminal is connected to the anode (P material). The negative polarity of voltage on the cathode repels the majority carriers (electrons) and causes them to move toward the junction. Likewise the positive polarity of voltage on the anode repels the majority carriers (holes) and forces them to-

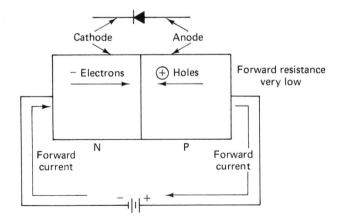

FIGURE 2–4   Forward-biased PN junction diode.

ward the junction. The majority carriers (electrons and holes) meet in and around the depletion area and cancel each other by joining or recombining. For each hole and electron that join together or recombine, an electron moves onto the cathode (N material) and off of the anode (P material) representing external current. Figure 2–5 illustrates the movement of the majority carriers when a diode is forward biased.

It should be noted that each time an electron moves, a hole is created in the space from which it moved. Likewise a hole is filled in the space to which it moved. Thus, electrons move and create holes which seem to "move" in the opposite direction. The minority current carriers of both materials are attracted by the external voltage polarity and recombine. The action of minority carriers causes a very small portion of external current.

REVERSE BIASING

A PN junction diode may also be reverse biased. When the anode of a diode is made negative and its cathode is made positive, it is said to be *reverse biased*. The reverse resistance of a reverse-biased diode is very very large, resulting in a very very small reverse current flow. Figure 2–6 illustrates a PN junction diode that is reverse biased.

The external current that results when a diode is reverse biased is so small that it is extremely difficult to measure. This is true because all reverse current is due to the action of the few minority carriers in the N and P materials.

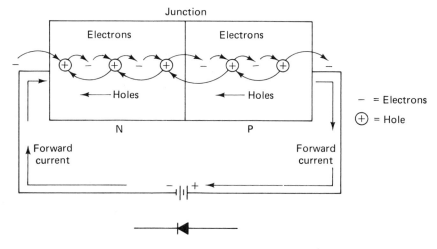

FIGURE 2–5 Movement of majority carriers with forward-biased diode.

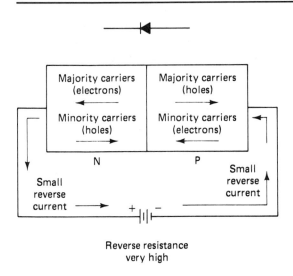

FIGURE 2–6   Reverse-biased PN junction diode.

When the cathode of a diode is made positive and the anode neg-ative, the electrons (majority carriers) in the N material are attracted away from the junction while the holes (minority carriers) are repelled toward the junction. Likewise the holes (majority carriers) of the P ma-terial are attracted away from the junction while the electrons (minority carriers) are repelled toward the junction. This action causes the few available minority carriers to join or recombine. For each recombination an electron moves onto the P material (anode) off of the N material (cathode).

Figure 2–7 illustrates typical diode characteristics. The curves on the graph represent the action of the diode when either forward or re-verse biased. From this illustration it can be seen that only a few tenths of a volt in the forward direction will cause the diode to reach the "turn-on" point. The turn-on point is the point at which the forward voltage overcomes the potential barrier and causes the diode to conduct. This is like a SPST switch being "turned on." Likewise, a reverse voltage of many volts causes very little conduction. This is like a SPST switch being "turned off."

However, as shown on the reverse conduction curve, there is a point in the operation of all diodes that is known as the zener, reverse breakdown, or reverse avalanche point. This point is a reverse voltage value that forces the diode to conduct large amounts of reverse current. This large reverse voltage causes the material's covalent bonds to rup-ture and usually results in the damage or destruction of the diode.

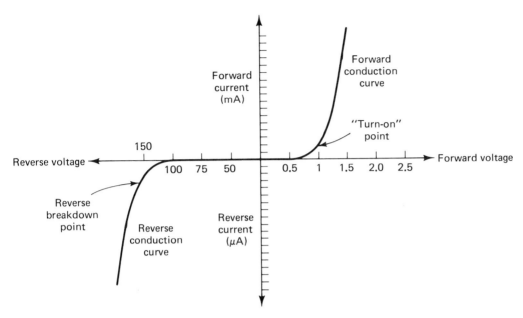

FIGURE 2–7   Typical diode characteristics.

PN JUNCTION DIODE RESISTANCE

It can be seen that a diode, when forward biased, exhibits a very low resistance. Likewise, a reverse-biased diode exhibits a very high resistance. These resistances can be very hard to measure with an ohmmeter insomuch as a change in the ohmmeter's scale brings about a change in the diode's resistance. However, an ohmmeter can be used to "check" the condition of a diode.

An ohmmeter has its own internal DC power supply; thus, one lead of the ohmmeter is negative while the other is positive, as illustrated in Figure 2–8. If the positive lead of an ohmmeter is placed on the anode of the diode to be "checked" and the negative lead is placed on the cathode, the diode would be forward biased. This will cause current to flow through the meter's movement, indicating a very low forward resistance as illustrated in Figure 2–9. If the range of the ohmmeter is changed, causing a higher internal meter voltage to be applied to the diode, the diode's forward resistance would decrease.

By reversing the ohmmeter's leads across the diode being checked, we reverse bias the diode. This should cause the ohmmeter to indicate

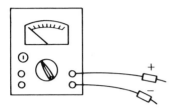

FIGURE 2–8   Typical ohmmeter.

a very high resistance, or even an infinite resistance in some cases. Figure 2–10 illustrates the ohmmeter being used to check a diode's reverse condition.

The ohmmeter can only be used to check the overall condition of any diode. If the diode in question exhibits a low resistance when forward biased by the ohmmeter and a high resistance when reverse biased by the meter, it is probably in good condition. If the ohmmeter indicates an unusually high or infinite forward resistance, the diode being checked might be "open" (a condition that would not allow very much, if any, forward current). If the ohmmeter indicates a low reverse resistance, the diode in question might be "shorted" (a condition that would allow current to flow in both directions through the diode). In either instance the diode's condition would be questionable, and the diode should be replaced.

The forward and reverse resistance may be accurately determined by using a diode's characteristic conduction curve. Figure 2–11 illustrates the forward conduction curve of a diode. Let us determine the forward resistance of a diode with the forward conduction characteristic illustrated in Figure 2–11. Assume that the diode has 0.8 V across it. Move along the voltage axis of the graph until you come to 0.8 V. At that point draw a vertical line upward until the diode's characteristic curve is reached. This is indicated as point X on the curve. From point X draw

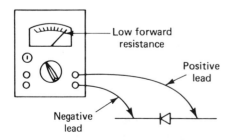

FIGURE 2–9   Ohmmeter with forward-biased diode.

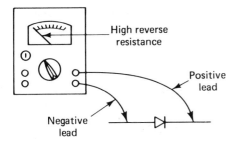

FIGURE 2–10   Ohmmeter with reverse-biased diode.

a horizontal line to the left until the current axis is intersected. This should indicate that 0.8 V across the diode results in 6 mA of forward current. The forward resistance of the diode at that point (X) is equal to 133.33 Ω and is found by applying Ohm's law:

$$R = \frac{E}{I} = \frac{0.8}{0.006} = 133.33 \ \Omega$$

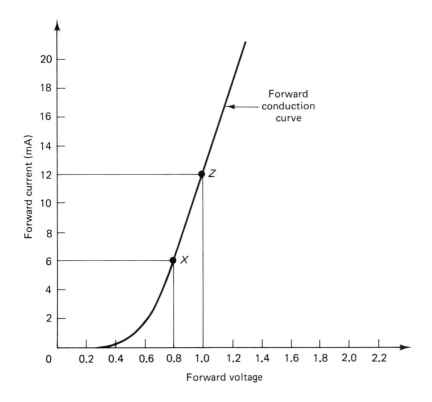

FIGURE 2–11   Forward conduction curve.

By applying the same procedure when 1.0 V appears across the diode, point Z on the characteristic curve equals 12 mA of forward current. Thus, the diode's forward resistance of 83.33 Ω is calculated as follows:

$$R = \frac{E}{I} = \frac{1.0}{0.012} = 83.33 \ \Omega$$

It can be seen that as a diode's forward voltage increases, its forward resistance decreases.

When a diode is connected into a circuit with an external load resistance, the characteristic curve of a diode can be used with an applied *load line* to determine important circuit values. Figure 2–12 illustrates the application of a load line. In this illustration a diode test circuit is represented by a diode, a 100-Ω resistor (R_1), and a DC power supply, all connected in series. To plot a load line for this circuit, find the power supply voltage on the forward voltage axis and label that as point A. We ignore the diode for the moment and apply Ohm's law to determine how much current will flow when 2 V (source voltage) is applied to 100 Ω of resistance (R_1) as follows:

$$I = \frac{E}{R} = \frac{2.0}{100} = 0.02 \ A \quad \text{or} \quad 20 \ mA$$

Now move along the current axis and establish point B at 20 mA. Draw a straight line between points A and B. This represents a 100-Ω load line and can be used to determine the current flow in the circuit, the voltage across the diode, the forward resistance of the diode, and the voltage across the resistor (R_1).

The point of intersection formed by the load line and the characteristic curve is known as the *operating point* and is labeled Q in our illustration. A straight line may be projected from point Q until that line intersects the voltage axis, to find that the voltage across the diode equals 0.8 V. Therefore, the voltage across R_1 equals 1.2 V (voltage across the diode minus the source voltage equals the remainder, or 2.0 − 0.8 = 1.2 V).

A straight line projected to the left from point Q until it intersects the current axis indicates that 11 mA of current flows in this circuit. Thus, the diode's forward resistance at this point equals 72.72 Ω and is found as follows:

$$R = \frac{E}{I} = \frac{0.8}{0.011} = 72.72 \ \Omega$$

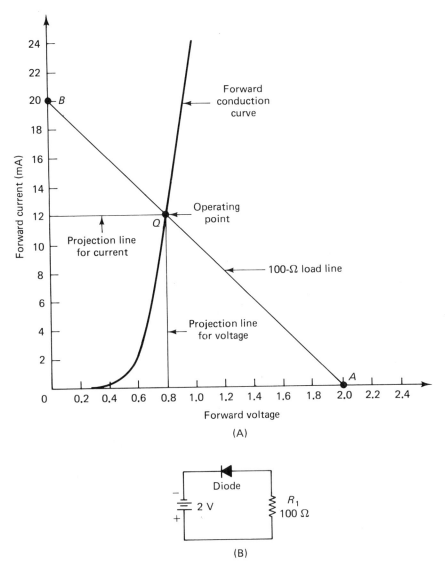

FIGURE 2–12 Diode load line. (A) Voltage versus current.
(B) Diode test circuit.

The total resistance of the circuit, omitting the resistance of the
power supply, equals 172.72 Ω (resistance of R_1 plus the forward resistance of the diode equals total resistance, or $100 + 72.72 = 172.72$ Ω). If an
AC voltage is substituted for the DC voltage in the test circuit, it can be
seen that the diode's forward resistance would instantaneously change
with the instantaneous AC peak voltage. As a diode's forward voltage
increases, its forward resistance decreases.

DIODE SPECIFICATIONS

Manufacturer specifications are available for all PN junction diodes. Generally these specifications include, among others, the maximum forward current and the maximum forward and reverse voltage associated with the diode.

The *maximum forward current* rating of a diode is the largest amount of current that can be allowed to pass through the diode while forward biased. The *maximum forward voltage* rating is the greatest amount of forward-bias voltage to be applied across the diode's terminals at any instant. The *maximum reverse voltage* is the largest amount of voltage that the diode can withstand when reverse biased and is generally called the *peak inverse voltage (PIV) rating*. All of these specifications are provided with the diode at a constant temperature of 25 °C.

The diode's operating temperature is of special interest. All semiconductors are said to exhibit a *negative temperature coefficient*. This is to say that as the temperature of a diode increases, its resistance will decrease. If the operating temperature of a diode is not carefully controlled, its internal resistance could decrease, causing an increase in circuit current. This can cause the diode to malfunction and do damage to itself or to other components. This condition is referred to as *thermal runaway* and must be closely guarded against.

Diodes may be broadly classified as signal or power diodes. Generally this classification is dependent upon their maximum forward current rating. A diode that has a maximum forward current rating listed in milliamperes is classified as a *signal diode*. Diodes designed to handle 1 A or more of forward current are said to be *power diodes*. Power diodes that can safely carry many many amperes of forward current have been developed and are in use today. PN junction diodes are found in all kinds of electronic circuits that rectify, detect, clip, clamp, count, and act as logic gates. These applications along with others will be examined in later chapters.

SPECIALIZED DIODES

There are several types of specialized diodes used today. These specialized diodes will be discussed next.

ZENER DIODE

The *zener diode* is carefully manufactured to be operated in a reverse-biased condition. It is doped in such a way as to allow its reverse breakdown voltage to be reached and exceeded during its normal operation, resulting in large amounts of reverse current. A normal PN junction diode would be destroyed if this was allowed to happen.

Figure 2–13 illustrates the characteristics of the zener diode. The forward conduction characteristics of the zener diode are identical to those of the PN junction diode. The reverse conduction characteristics are very different and will be examined next.

The reverse breakdown voltages of zener diodes range from a few volts to hundreds of volts. When this breakdown voltage across the zener diode is reached or exceeded, its internal resistance is lowered and a large reverse current flows. A unique characteristic of this diode's action is that once it breaks down, its internal resistance decreases with each increase in reverse voltage, causing the voltage drop across the ze-

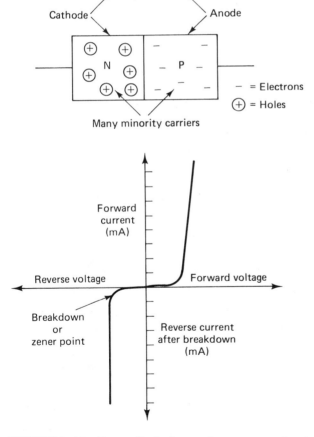

FIGURE 2–13   Zener diode forward-reverse conduction characteristics.

ner to remain constant. This constant voltage drop across the zener and the zener's breakdown voltage are one and the same. Figure 2–14 illustrates this characteristic. Figure 2–14 shows the characteristics of a zener diode in three separate circuits. The reverse conduction curve illustrates that its breakdown voltage is 10 V. This is to say that when 10 V or more are applied to this zener, reverse current flows and 10 V will remain constant across the zener.

In all three circuits the value of R_1 is the same. The same 10-V zener is used in all three circuits. Likewise in all three circuits the voltmeter measures a constant 10 V even though in circuit 1 the input voltage is 15 V, in circuit 2 the input voltage is 20 V, and in circuit 3 the input voltage is 25 V.

In order to cause the zener to remain "on," a very small reverse current must be allowed to flow through it at all times. If this small current is not maintained, the zener will be turned "off." This small reverse current is called its *"holding" current* and is illustrated in Figure 2–14.

The two most important ratings associated with a zener diode are its voltage and power ratings. The zener's *voltage rating* indicates the voltage at which the device is caused to conduct as well as the constant voltage drop maintained across the zener during conduction. The *power*

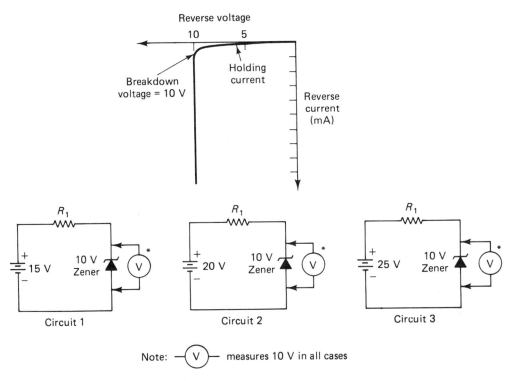

FIGURE 2–14   Zener diode's breakdown voltage and holding current. Note: —Ⓥ— measures 10 V in all cases.

rating (listed in watts) indicates the maximum reverse current that a zener can conduct without damage. Thus a 10-V, 1-W zener could safely conduct 0.1 A, or 100 mA, of reverse current as calculated below:

$$I = \frac{P}{E} = \frac{1}{10} = 0.1 \text{ A} \quad \text{or} \quad 100 \text{ mA}$$

The zener diode is used primarily as a voltage regulator. Its applications will be examined in later chapters.

VARICAP (VARACTOR) DIODES

The varicap diode is a device that when reverse biased acts like a small variable capacitor. Recall that capacitance exists any time two areas of unlike charge are separated by a dielectric (nonconductor). This is to say that a small amount of capacitance would exist if two small metal plates separated by air (an excellent dielectric) were given unlike charges, as illustrated in Figure 2–15. In this illustration a small capacitance (measured in picofarads) would exist between the metal plates. This capacitance would be determined by the area of the plates, the distance between the plates, and K, the dielectric constant.

Recall the existence of the depletion zone in PN junction diodes. This depletion zone becomes wider when the PN junction is reverse biased and narrower when forward biased. Since there are no current carriers in the depletion zone, this zone serves as a very good dielectric. The P and N materials on either side of the depletion zone become the areas of unlike charges.

Thus, the varicap is specifically manufactured to cause the width of the depletion zone to be altered by a reverse-bias voltage. Figure 2–16 serves to illustrate the action of the varicap diode. As the distance between the plates of a variable capacitor increases, capacitance is decreased. Figure 2–16 illustrates that as the reverse voltage increases, the capacitance between the N and P material decreases due to the width of depletion zone increasing. Thus, the varicap diode serves as a small variable capacitor whose capacitance is varied from a few picofarads to near 100 pF by reverse voltage. This device is used as a frequency control in communication electronics.

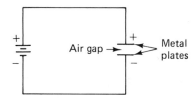

FIGURE 2–15 Characteristics of capacitance.

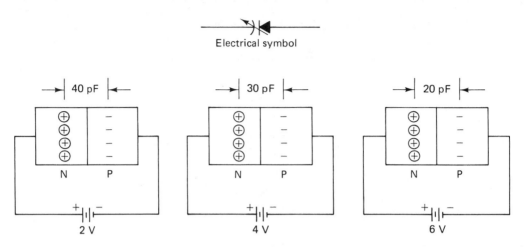

FIGURE 2–16   Varicap diode and capacitance.

TUNNEL DIODES

Tunnel diodes have very specialized applications due to the unusual characteristics demonstrated when they are forward biased. Recall that all conductors of electric current exhibit resistance. This resistance may be classified as either positive or negative resistance. Most conductors exhibit a positive resistance. That is, as the voltage increases, the current will increase accordingly. Some conductors exhibit a negative resistance. A conductor is said to exhibit negative resistance when as voltage is increased, current actually decreases. The normal PN junction diode exhibits a positive resistance when forward biased. A tunnel diode exhibits both a negative and positive resistance when forward biased.

Figure 2–17 illustrates the characteristics of the tunnel diode. The tunnel diode, sometimes called the *Esaki diode* for its Japanese inventor, is heavily doped, resulting in the formation of a very thin depletion zone. Because of this very thin depletion zone, a few tenths of forward voltage will cause current carriers to be forced on (tunneled through) the depletion area, resulting in immediate conduction. This conduction is represented between points *A* and *B* on the characteristics curve illustrated in Figure 2–17. This immediate movement of carriers serves to build up the potential barrier, resulting in a decrease of current flow from points *B* to *C*. From point *C* the diode exhibits the usual forward conduction characteristics of a PN junction diode. Thus, from points *A* to *B* on the conduction curve the tunnel diode exhibits a positive resistance. From points *B* to *C* this device exhibits a negative resistance. Beyond point *C* the positive resistance characteristic prevails. The tunnel diode is used in oscillator circuits.

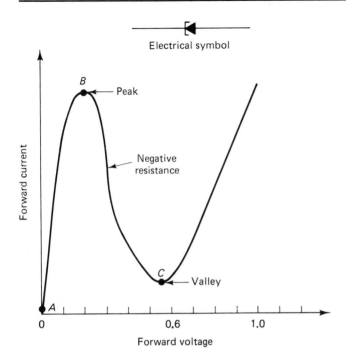

FIGURE 2–17   Characteristics of the tunnel diode.

It should be noted at this point that the condition of the tunnel diode should never be checked with an ohmmeter. This action could result in damage to the diode because of its very low power ratings.

VACUUM DIODES

Vacuum diodes are electron tubes that have been generally replaced by semiconductor diodes. This is true because semiconductor diodes are more reliable, cheaper, last longer, are smaller, are more rugged, and do not rely upon heat for their operation as do many electron tubes.

Even though the state of the art is reflected by the use of semiconductors, there are several vacuum diodes still being used in the communications and manufacturing industries. For this reason, we will describe their characteristics in this chapter. Recall from Chapter 1 that Edison discovered how current flowed in a vacuum. J. A. Fleming invented what was known as the Fleming "valve," a two-electrode device that was the forerunner of the modern vacuum diode tube.

Vacuum diodes may be directly or indirectly heated. Figure 2–18 illustrates symbols for both types of tubes along with the elements (internal conductors) contained within each tube. Most modern-day vac-

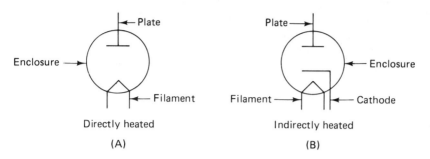

FIGURE 2–18 Directly-indirectly heated vacuum tubes. (A) Directly heated. (B) Indirectly heated.

uum diodes are indirectly heated and are enclosed by glass, although some vacuum diodes employ metal enclosures.

In Chapter 1 we discussed how current flows in a vacuum. A voltage is applied to the filament of the directly heated vacuum tube causing it to be heated to incandescence and to emit electrons. These electrons are attracted to the positively charged plate. Likewise a voltage must be applied to the filament of the indirectly heated tube. The filament of the indirectly heated tube does not emit electrons. An additional element known as the *cathode* has been placed around the filament. The cathode, sometimes called the *emitter*, emits electrons that are attracted to the positively charged plate. In both tube types the space cloud is formed around the filament of the directly heated tube and around the cathode of the indirectly heated type.

Figure 2–19 illustrates the typical construction of the indirectly heated vacuum diode along with a typical circuit. In some instances enough elements have been enclosed in one glass enclosure to cause the electron tube to act as two or even three tubes. These types of vacuum diodes are known as dual and triple diodes. Typical diagrams of these tubes are illustrated in Figure 2–20.

Vacuum diode tubes can have as many as nine external leads affixed to internal elements. The external leads are fastened to pins that are numbered for identification purposes. The tube is usually plugged into a tube socket that has holes that are also numbered for identification purposes. A tube manual must be used to determine the numbered pin that corresponds to the tube's internal element in order to properly connect the tube into a circuit. Figure 2–21 illustrates a typical tube and socket along with a typical base diagram taken from a tube manual.

The tube would be plugged into the socket and the numbered elements would be used to connect the circuit illustrated in Figure 2–22. Notice that it is essential that the appropriate pin/hole must be connected to the correct component and/or power source if the circuit is to function properly.

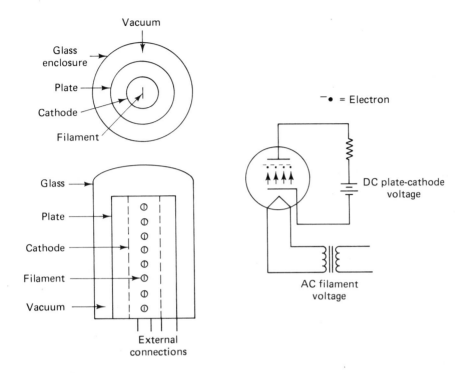

FIGURE 2–19   Typical vacuum tube construction and circuit.

When examining the characteristics of a vacuum diode, plate current, plate voltage, and plate resistance are considered. The *plate voltage* is the voltage drop across the tube or between the plate and cathode during any portion of its operation. The *plate current* is the current flow that takes place between the cathode and plate. The *plate resistance* is the internal resistance of the tube when it is conducting current. It should be noted that for current to flow through a vacuum diode, the

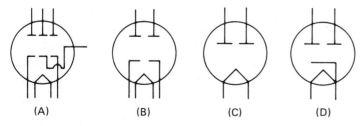

FIGURE 2–20   Typical vacuum tubes. (A) 6GQ7. (B) 6AL5.
(C) 5X3GT. (D) 5V4G.

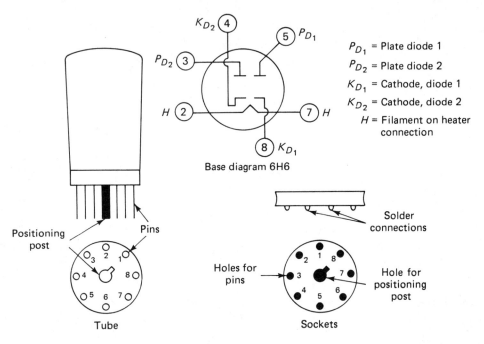

P_{D_1} = Plate diode 1
P_{D_2} = Plate diode 2
K_{D_1} = Cathode, diode 1
K_{D_2} = Cathode, diode 2
H = Filament on heater connection

Base diagram 6H6

FIGURE 2–21   Typical tube and base.

plate must always be positive when compared to the cathode. Current will never flow from the plate to the cathode unless of course the tube is damaged.

Figure 2–23 illustrates a typical characteristic conduction curve of a vacuum diode. It should be noted that the vacuum diode does not perform as well as the PN junction semiconductor diode. This is due to the tube having a much higher internal resistance along with other factors.

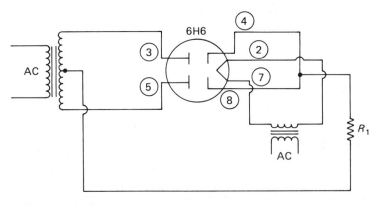

FIGURE 2–22   Typical vacuum tube circuit.

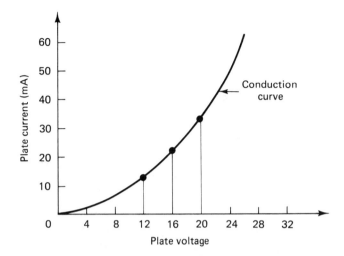

FIGURE 2–23   Plate resistance and calculations.

Much higher voltages are normally associated with the operation of the vacuum diode than those of the PN junction diode. The curve illustrated in Figure 2–23 may be used to compute plate resistance. These values are usually very high. For example, the plate resistance when the plate voltage is 12, 16, and 20 V may be calculated. A line is projected from each of these points until it intersects the conduction curve, and across until it intersects the plate current axis. The plate resistance for each point of reference is thus determined. When the plate voltage is 12 V, the plate resistance is 1000 Ω as shown below:

$$R = \frac{E}{I} = \frac{12 \text{ V}}{0.012 \text{ A}} = 1000 \text{ Ω}$$

Likewise the plate resistance is 727.2 and 666.6 Ω when the plate voltage is 16 and 20 V, respectively, as shown below:

$$R = \frac{E}{I} = \frac{16 \text{ V}}{0.022 \text{ A}} = 727.2 \text{ Ω} \qquad R = \frac{E}{I} = \frac{20 \text{ V}}{0.030 \text{ A}} = 666.6 \text{ Ω}$$

It can be seen that as plate voltage increases, plate resistance decreases.

The conduction curve of a diode may be used to determine its characteristics while in a loaded condition (external resistance connected to its plate). Examine the conduction curve and circuit illustrated in Figure 2–24. If a load line is used with this conduction curve, circuit values

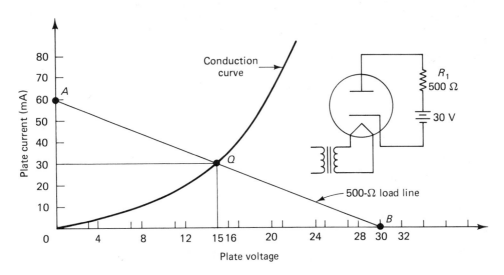

FIGURE 2–24 Load line characteristics with circuit.

may be determined. Assume the tube is not in the circuit. Compute the amount of current flow that would result when 30 V (power supply voltage) is applied to 500 Ω (the resistance in the circuit). This equals 60 mA ($I = E/R = 30$ V/500 Ω $= 0.06$ A, or 60 mA). Find 60 mA on the plate current axis and note that it is labeled point A. Now we determine the voltage across the tube if the resistor was not in the circuit. That voltage would equal the power supply voltage, which is 30 V. Now find 30 V on the plate voltage axis and note that it is labeled point B. Points A and B are connected by a straight line. This line represents a 500-Ω *load line*. The point of intersection formed by the load line and conduction curve establishes important circuit values such as plate current, plate resistance, plate voltage, and voltage drop across the resistor.

By projecting a line from point Q to the plate current axis, the plate current is determined to be 30 mA. Another line projected from point Q to the plate voltage axis indicates that the plate voltage is 15 V. The voltage across the resistor is 15 V also. (Source voltage minus plate voltage equals the remainder, or $30 - 15 = 15$ V.) Plate resistance is 500 Ω ($R = E/I = 15$ V/0.030 A $= 500$ Ω). If additional load lines representing both larger and smaller values of load resistance are drawn, it becomes obvious that as plate voltage increases, plate resistance will decrease. If an AC voltage was substituted for the DC voltage in the example, it can be seen that the diode's resistance would instantaneously change with the instantaneous AC peak voltage. AC average values of resistance may then be considered.

Vacuum Diode Specifications

Some of the significant manufacturer's specifications to consider when selecting a vacuum diode for a specific circuit are the maximum plate current, plate voltage, reverse voltage (peak inverse voltage—PIV), and the filament voltage. The *maximum plate current* is the largest amount of current that can be allowed to pass between the cathode and plate without causing damage to the tube. The *maximum plate voltage* is the largest amount of voltage that can exist between the plate and cathode without destroying the tube's elements. The *PIV* is the maximum amount of reverse voltage the tube can withstand (cathode positive, plate negative) before being damaged. The *filament or heater voltage* is the manufacturer's recommended voltage (AC or DC) to be applied to the heater of the tube to cause it to operate efficiently.

The vacuum diode tube is generally used as a half- or full-wave rectifier in power supply circuits; however, it is rapidly becoming obsolete.

HOT-CATHODE GASEOUS DIODES

In Chapter 1 current flow in a gas was discussed. Current flow through the hot-cathode gaseous diode is due to the emission of electrons causing ionization of the gas contained in the glass enclosure. Figure 2–25 illustrates the symbol for the hot-cathode gaseous diode. Since some type of inert gas (mercury vapor, argon, neon) has been placed within the enclosure, a dot is used in the symbol to indicate the presence of a gas.

When an external voltage is applied to the filament of the diode, current flows through the filament causing it to heat up and emit electrons. These electrons form a negative space cloud around the filament of the tube. When the plate is made positive as compared to the filament (emitter), electrons are attracted by, and move toward, the plate. The electrons strike the gas molecules within the enclosure, knocking additional electrons from the gas molecules. When a gas molecule loses an

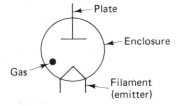

FIGURE 2–25   Symbol of hot-cathode gaseous tube.

electron, it becomes a positive ion and is attracted toward the negative filament. The electrons that are knocked from the gas molecule join the electrons emitted by the filament on their journey toward the plate. When the plate becomes positive enough to attract a sufficient number of electrons, causing a sufficient number of ions to be formed, the gas is *ionized*. The ionization of the gas results in a large current flow through the tube, limited only by an external load. The minimum voltage between the plate and filament that causes ionization of the gas is known as the *"firing" potential* and is slightly greater than the ionization potential. The *ionization potential* for most tubes is determined by the physical characteristics of the tube as well as the type of gas placed in the tube. For example, the ionization potential for mercury vapor is near 10 V, while helium's ionization potential is near 24 V. Figure 2–26 illustrates a typical characteristic conduction curve of a hot-cathode gaseous diode, along with a simple circuit.

Generally, the gaseous diode is a better diode than the vacuum diode as demonstrated by its linear conduction once ionization takes place. Additionally, the voltage across the tube remains nearly the same as current through the tube increases. As indicated by the "knee" portion of the conduction curve, the "firing" potential of the tube is actually slightly greater than the ionization potential of the gas.

When using this type of tube in a circuit, special precautions must be observed. A "warm-up" time of several seconds must be allowed before a positive voltage is applied to the tube's plate. The voltage that causes the filament to heat up must be applied several seconds before the plate voltage is applied to allow a sufficient space cloud to be formed around the filament. If a positive voltage is applied to the plate before a space cloud is formed around the filament, premature ionization would cause the filament to be bombarded by heavy positive ions resulting in damage to the tube.

The hot-cathode gaseous diode is never operated without some external load. Unlike the vacuum diode that exhibits a high internal resistance, the gaseous diode's internal resistance drops to a very low level once ionization takes place. This results in a very large external current that must be limited by an external load. Without an external load, this tube's current would rise to a level that would damage or even melt the tube's elements.

Like the vacuum tube, the external connections that represent the internal elements of a gaseous diode must be identified by using an appropriate tube manual. Specifications of interest include filament voltage, firing potential, maximum plate current, and PIV. Like the vacuum tube, the gaseous diode only conducts in one direction. Current always flows from the filament or emitter to the plate.

The hot-cathode gaseous diode is typically used for high voltage-current rectification in power supply circuits.

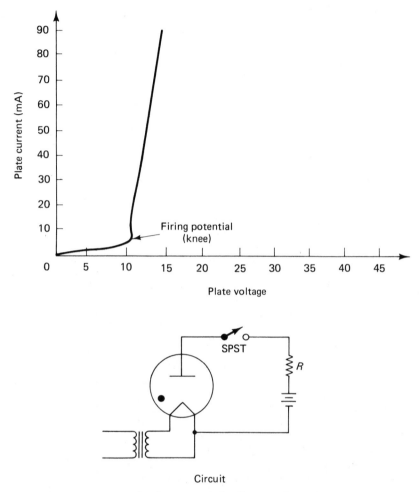

FIGURE 2–26   Hot-cathode gaseous diode's conduction curve.

COLD-CATHODE GASEOUS DIODES

The cold-cathode diode is a gas tube that relies on a high voltage between its plate and cathode to cause ionization of its gas. This tube is also known as a *glow discharge tube* and does not contain a heater or filament. Figure 2–27 illustrates the symbol and elements of this tube, as well as a characteristic conduction curve. Like the vacuum and hot-cathode gaseous diode, the cold-cathode tube conducts in one direction. Current always flows from the cathode to the plate.

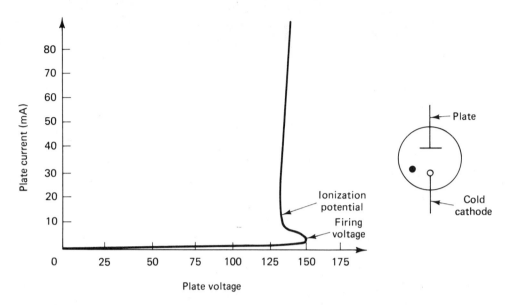

FIGURE 2–27   Cold-cathode gaseous tube.

The conduction curve indicates that a rather large voltage must exist between the plate and cathode of this tube to "turn it on." Also, the firing potential is several volts higher than the ionization potential.

As the plate of this tube is made more positive, the free electrons that exist in limited numbers in all gases move toward the plate. Likewise positive ions move toward the cathode. This effect is cumulative. As the plate voltage rises, additional electrons and positive ions are affected. Their movement generates additional electrons due to the "knock-off" process. When the plate voltage is great enough, the gas ionizes and the tube breaks into conduction. Its internal current is limited only by an external load. The most significant characteristic regarding this type of diode is that once ionization takes place, the voltage drop across the tube remains constant. This characteristic causes the cold-cathode diode to act as a very good *voltage regulator*.

Like the vacuum and hot-cathode gaseous diode tubes, the cold-cathode diode employs a pin and socket arrangement for circuit connection. The internal tube elements along with the corresponding external connections may be identified by using an appropriate tube manual. It is interesting to note that some gaseous diode tubes employ an external anode or plate cap rather than a pin connection. This arrangement provides the means to connect the plate of the tube in a circuit and is illustrated in Figure 2–28.

In this chapter the characteristics of diode devices were examined. PN junction diodes were discussed along with the zener, tunnel, vari-

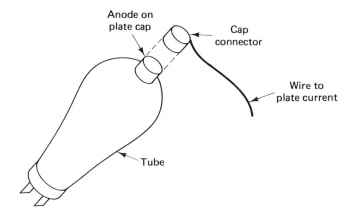

FIGURE 2–28 Tube with plate cap.

cap, vacuum, and gaseous diodes. Answer the following review questions, solve the problems, and perform the indicated laboratory activities to supplement your understanding of diode devices. These activities will help to develop a further understanding of the characteristics of diode devices.

REVIEW

1. What are the electrical symbols for the PN junction, zener, tunnel, and varicap diodes?

2. What type of semiconductor materials form the anode and cathode of a diode?

3. Why does the depletion area of a PN junction exist?

4. Why is the potential barrier of a PN junction diode critical to its conduction?

5. What is meant by forward and reverse biasing?

6. How does a diode's forward resistance compare to its reverse resistance?

7. What is likely to happen when the reverse breakdown voltage of a normal PN junction diode is exceeded?

8. How can a PN junction diode's condition be checked with an ohmmeter?

9.   What can be determined by applying a load line to a diode's forward characteristic curve?

10.   What is a negative temperature coefficient?

11.   What manufacturer's specifications should be considered when selecting a diode for use in a circuit?

12.   How does the zener diode differ from the PN junction diode?

13.   What two manufacturer's specifications must be considered when selecting a zener diode for use in a circuit?

14.   How do the zener's voltage rating, its reverse breakdown voltage, and the voltage drop across a zener compare?

15.   How does a varicap diode act as a variable capacitor?

16.   What is the relation between reverse voltage and the capacitance of a varicap diode?

17.   Why is a tunnel diode said to exhibit negative resistance?

18.   What are the differences between a directly and indirectly heated vacuum diode?

19.   How are vacuum diodes connected in electronic circuits?

20.   What necessary information is provided by an appropriate tube manual relative to connecting a vacuum diode into an electronic circuit?

21.   What is the relation between the plate voltage and plate resistance of a vacuum diode?

22.   What information may be determined by applying a load line to the characteristic conduction curve of a vacuum diode?

23.   What manufacturer's specifications are important when selecting a vacuum diode for a specific circuit?

24.   What are the firing and ionization potentials of a hot-cathode gaseous diode and how do they differ?

25.   What are the differences between the hot-cathode and cold-cathode gaseous diodes?

26.   Why must a sufficient "warm-up" time be provided for the operation of the hot-cathode gaseous diode?

27.   Why must both the hot- and cold-cathode gaseous diodes be operated with an external load?

28.   What factors determine the ionization potential of a gaseous diode?

29.   What causes the gas within a gaseous diode to ionize?

PROBLEMS

1. Using the forward conduction curve illustrated in Figure 2–11, compute the diode's resistance when its forward voltage equals 0.7, 0.9, and 1.1 V.

2. Using the forward conduction curve illustrated in Figure 2–12, draw a load line and compute the circuit current, voltage across the diode, voltage across R_1, diode resistance, and total circuit resistance for the circuit in Figure 2–29.

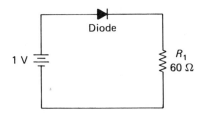

FIGURE 2–29   PN diode calculations.

3. Compute the maximum reverse current for the zener in Figure 2–30.

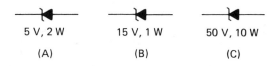

| 5 V, 2 W | 15 V, 1 W | 50 V, 10 W |
| (A) | (B) | (C) |

FIGURE 2–30   Zener diode calculations. (A) 5 V, 2 W. (B) 15 V, 1 W. (C) 50 V, 10 W.

4. Using the characteristic conduction curve found in Figure 2–23, compute the tube's plate resistance when its plate voltage equals 14, 18, and 22 V.

5. Using the characteristic curve and circuit illustrated in Figure 2–24 and appropriate load lines, determine the plate voltage, plate current, voltage across R_1, and plate resistance when the value of R_1 equals 750, 1000, and 1500 Ω.

SUGGESTED LABORATORY ACTIVITIES

1. Identify and label the anode and cathode leads of several PN junction diodes with an ohmmeter.

2. Identify the shorted and open devices from a group of several PN junction diodes (some of which are open or shorted) with an ohmmeter.

3. Using appropriate values of components, construct the circuits in Figure 2–31 and measure the total current.

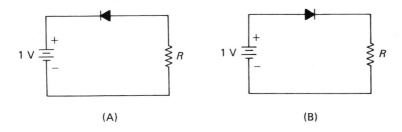

(A)   (B)

FIGURE 2–31   PN circuits for construction.

4. Using three different values of zener diodes and the appropriate values of resistance and power supply voltages, construct the circuits in Figure 2–32 and measure and record the zener's reverse breakdown voltages.

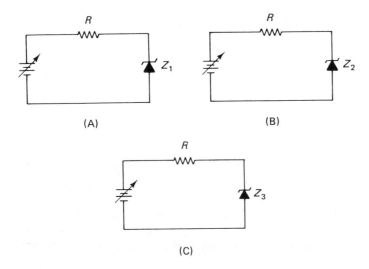

(A)   (B)

(C)

FIGURE 2–32   Zener circuits for construction.

5. Using an appropriate tube manual as reference, draw the base diagrams for the following tubes: 5V3A, 6AF3, 12BT3, 866A, 0A2.

6. Construct the circuit in Figure 2–33. Adjust the power supply at 2-V intervals from 0 to 20 V and record the plate current for each plate voltage. Using this data, construct a characteristic conduction curve on the graph in Figure 2–34.

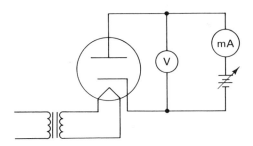

FIGURE 2–33   Tube circuit for construction.

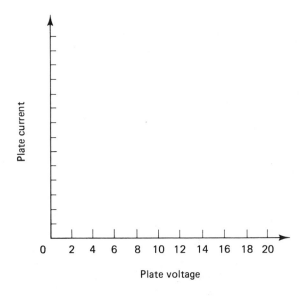

FIGURE 2–34   Grid for conduction curve.

7. Construct the circuits in Figure 2–35 and determine the firing and ionization potentials for each tube.

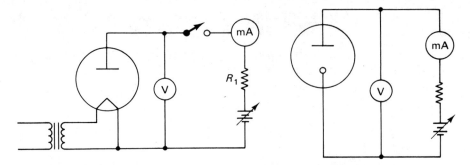

FIGURE 2–35   Tube characteristics circuit.

CHAPTER 3
Diode
Applications—
Power Supply
Circuits

Power supply circuits are commonly used in electronic equipment. The purpose of a power supply circuit is to deliver the proper values of voltage and current to the various circuits of electronic equipment. In some equipment, such as portable radios and calculators, the power supply is a battery or dry cell. Equipment such as televisions, stereos, and many other electronic instruments use AC supplies as their energy source. A power supply circuit for these types of equipment converts AC energy into direct current, which is delivered to the circuitry within the equipment. Proper DC voltages and current levels cause the equipment to function properly.

Most of the electric power produced at power plants is 60-Hz, three-phase alternating current. However, the use of direct current is necessary for many types of equipment. Batteries and DC generators are sometimes used as DC energy sources. However, in many cases, it is more economical to use a circuit for converting alternating current into direct current. Direct current may be produced by circuits called *rectifiers*, which change alternating current into a form of direct current. Some types of rectifiers include (1) semiconductor rectifiers, (2) vacuum tube rectifiers, and (3) gaseous tube rectifiers. Semiconductor rectifiers are used much more extensively than the other types. Power supply circuits are common circuits that utilize electronic diode devices.

Another method of supplying the necessary operating voltages to electronic equipment is a *voltage divider circuit*. Voltage dividers may be used in power supplies when direct current is used as the voltage source. Voltage divider circuits are also discussed in this chapter since they are used in power supplies of electronic equipment.

CIRCUITS

Before discussing types of power supply circuits and various other circuit applications, it is necessary to look at methods of defining circuits. A *circuit* may simply be defined as a path or conductor through which electric current flows. Electric current flows only when it has a complete or closed circuit path. There must be a *source* of electric energy to cause current to flow along a closed path. Work is performed as electric current flows through a closed circuit. Electric energy produced by the source is converted to another form of energy in a closed circuit. *Closed circuits* have a certain amount of resistance to current flow, based on the values of the components in the circuit.

Electric current cannot flow if a circuit is open, since an *open circuit* does not provide a complete path for current flow. An example of an open circuit is when a light bulb "burns out." An open filament, in this case, stops current flow from the source. An open circuit therefore has infinite resistance to current flow.

Another type of circuit is the *short circuit*. A short circuit occurs when a conductor connects directly across the terminals of a source. If wire is placed across a battery, a short circuit occurs. Short circuits cause excessive current to flow from the source. Short circuits have nearly *zero resistance* to current flow.

Electronic circuits may be defined more easily by referring to Figure 3–1. A circuit is a part of some type of electronic system, such as a stereo. All systems have several subsystems, such as amplifiers, that cause the system to operate. An amplifier subsystem is composed of several circuits. A circuit has a source of energy from within the system and a load. The *load* converts the energy supplied to the circuits into another form, such as heat energy. Resistance in a circuit acts as a load since it converts electrical energy into heat energy. The speakers in a stereo system convert electric energy into sound energy. All circuits, such as power supplies, are composed of *components*, such as resistors, capacitors, and inductors. Circuits also have an input, the source of energy, and an output, which is usually derived from the load.

SINGLE-PHASE RECTIFICATION

The simplest circuit for converting alternating current to direct current is a single-phase rectifier. A *single-phase rectifier* changes alternating current to *pulsating* direct current. This process can be accomplished by

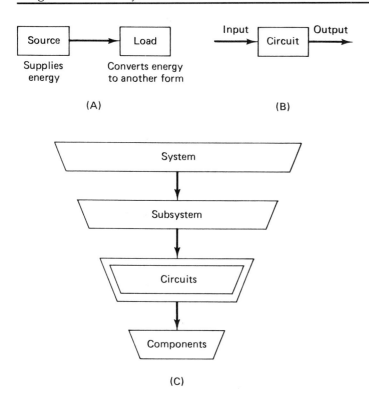

FIGURE 3–1 Circuits. (A) Circuits have a *source* and a *load*.
(B) Circuits have an *input* and an *output*. (C) Circuits are a part
of any *electronic system*.

using semiconductors, vacuum tubes, or gaseous tubes. The most common and economical method is the use of low-cost silicon semiconductor rectifiers.

SINGLE-PHASE, HALF-WAVE RECTIFIERS

A single-phase, half-wave rectifier circuit, shown in Figure 3–2, converts alternating current into pulsating direct current. Assume that during the positive alternation of the AC input cycle, the anode of the diode is positive. See Figure 3–2A. The diode will then conduct since it is forward biased and the junction resistance is low. The positive half-cycle of the AC input will then cause a voltage drop across the load device (represented by a resistor).

During the negative alternation of the AC input cycle, the anode of the diode becomes negative (see Figure 3–2B). The diode is now reverse biased and no significant current flows through the load device. There-

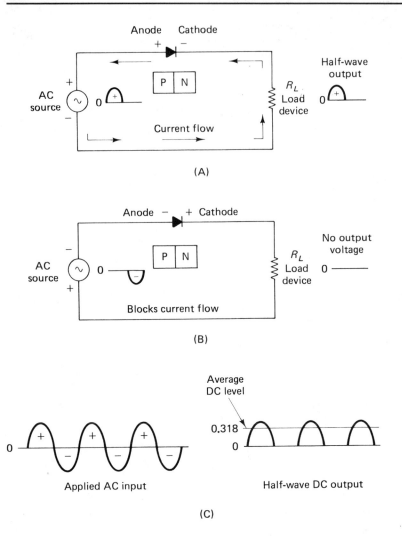

FIGURE 3–2   Single-phase, half-wave rectifier. (A) Forward-biased diode. (B) Reverse-biased diode. (C) Input-output voltage waveforms.

fore, there will be no voltage appearing across the load. The input and resulting output waveforms of the half-wave rectifier circuit are shown in Figure 3–2C. The pulsating direct current of the output has an average DC level. The average value of the pulsating direct current produced by single-phase, half-wave rectification is expressed as

$$V_{DC} = 0.318 \times V_{max}$$

where:

V_{DC} is the average DC value of rectified voltage and

V_{max} is the peak (maximum) value of applied AC input voltage.

Certain diode ratings should be considered for half-wave rectifier circuits. The *maximum forward current (I_{max})* is the largest current that can flow through the diode while it is forward biased without damaging the device. The *peak inverse (reverse) voltage (PIV)* is the maximum voltage across the diode while it is reverse biased. For the half-wave rectifier the maximum voltage developed across the diode is V_{max} of the applied AC input. As shown in the preceding formula, the PIV of a diode used in a half-wave circuit must be much larger than the DC voltage developed.

SINGLE-PHASE, FULL-WAVE RECTIFIERS

In order to obtain a purer form of DC energy, we must improve upon half-wave rectifier circuits. Figure 3–3 shows a single-phase, full-wave rectifier circuit that uses two diodes and produces a DC output voltage during each alternation of the AC input. The rectified output of the full-wave rectifier has twice the DC voltage level of the half-wave rectifier.

The full-wave rectifier utilizes a *center-tapped* transformer to transfer AC source voltage to the diode rectifier circuit. The instantaneous polarities of the transformer secondary during the positive half-cycle of AC source voltage are shown Figure 3–3A. The peak voltage (V_{max}) is developed across the transformer secondary. At this time, diode D_1 is forward biased and diode D_2 is reverse biased. Therefore, current flow occurs from the center tap, through the load device, through D_1, and back to the top terminal of the transformer secondary. The positive half-cycle of the AC input is developed across the load as shown.

During the negative half-cycle of the AC source voltage, diode D_1 is reverse biased and diode D_2 is forward biased. The instantaneous polarities are shown in Figure 3–3B. The current path is from the center tap, through the load device, through D_2, and back to the bottom terminal of the transformer secondary. The negative half-cycle of AC input is also produced across the load, developing a full-wave output as illustrated in Figure 3–3C.

Each diode in a full-wave rectifier circuit must have a PIV rating of twice the value of the peak voltage developed at the output. The average voltage output for a full-wave rectifier circuit is

$$V_{DC} = 2(0.318 \times V_{max})$$
$$= 0.636 \times V_{max}$$

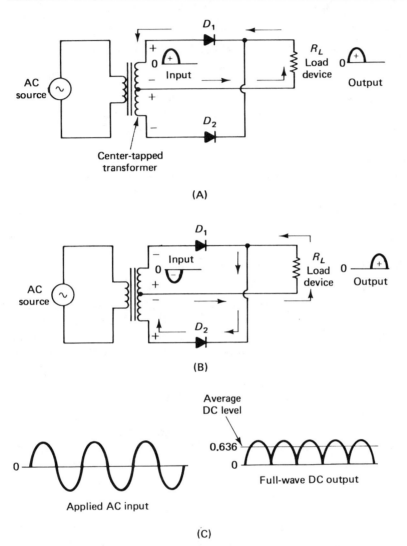

FIGURE 3–3 Single-phase, full-wave rectifier. (A) Diode D_1 forward biased. (B) Diode D_2 forward biased. (C) Input-output voltage waveforms.

This type of rectifier circuit produces twice the DC voltage output of a half-wave rectifier circuit. However, it requires a bulky center-tapped transformer as well as diodes that have a PIV rating of twice the peak value of applied AC voltage.

SINGLE-PHASE BRIDGE RECTIFIERS

One disadvantage of the full-wave rectifier discussed previously is the requirement of a large center-tapped transformer. To overcome this

disadvantage, four diodes can be used to form a full-wave bridge rectifier as shown in Figure 3–4. In addition, the diode PIV rating is only required to be at least the peak voltage value (V_{max}).

During the operation of a bridge rectifier, two diodes are forward
biased during each alternation of the AC input. When the positive half-
cycle occurs, as shown in Figure 3–4A, diodes D_1 and D_3 are forward
biased, while diodes D_2 and D_4 are reversed biased. This biasing condition is due to the instantaneous polarities occurring during the positive

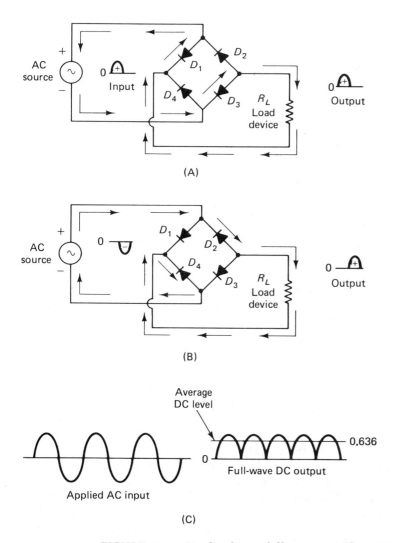

FIGURE 3–4   Single-phase, full-wave rectifier. (A) Diodes D_1
and D_3 forward biased. (B) Diodes D_2 and D_4 forward biased.
(C) Input-output voltage waveforms.

alternation. The conduction path is from the instantaneous negative side of the AC source, through diode D_3, through the load device, through diode D_1, and back to the instantaneous positive side of the AC source.

During the negative alternation of the AC input, diodes D_2 and D_4 are forward biased, while diodes D_1 and D_3 are reverse biased. Conduction occurs, as shown in Figure 3–4B, from the instantaneous negative side of the source, through D_2, through the load device, through D_4, and back to the instantaneous positive side of the AC source. Since a voltage is developed across the load device during both half-cycles of the AC input, a full-wave output is produced, as shown in Figure 3–4C, that is similar to that of the full-wave rectifier discussed previously. For high values of DC output voltage, the use of a bridge rectifier is desirable since the diode PIV rating is one-half that of the other single phase, full-wave rectifier method. Bridge rectifiers are now the most common types of rectifier circuits.

THREE-PHASE RECTIFICATION

THREE-PHASE, HALF-WAVE RECTIFIERS

Most industries are supplied with three-phase alternating current. Therefore, due to the inherent advantages of three-phase power, it is beneficial to use three-phase rectifier circuits to supply DC voltage for industrial use. Single-phase rectifier circuits are ordinarily used where low amounts of DC power are required. To supply large amounts of DC power for industrial requirements, a three-phase rectifier circuit such as the one shown in Figure 3–5 could be employed. Three-phase rectifier circuits produce a purer DC voltage output than single-phase rectifier circuits do, thus wasting less electric energy.

Figure 3–5 shows a three-phase, half-wave rectifier circuit that does not use a transformer. Phases A, B, and C of the wye-connected, three-phase source supply voltage to the anodes of diodes D_1, D_2, and D_3. The load device is connected between the cathodes of the diodes and the neutral of the wye-connected source. When phase A is at its peak positive value, maximum conduction occurs through diode D_1, since it is forward biased. No conduction occurs through D_1 during the negative alternation of phase A. The other diodes operate in a similar manner, conducting during the positive AC input alternation and not conducting during the associated negative AC alternation. In a sense, this circuit combines three single-phase, half-wave rectifiers to produce a half-wave DC output, as shown in Figure 3–5B. Of course, the voltages appearing across the diodes are 120° out of phase. There is a period of time during each AC input cycle when the positive alternations overlap one another,

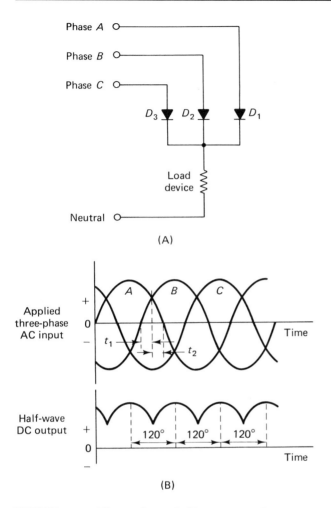

FIGURE 3–5   Three-phase, half-wave rectifier. (A) Schematic
diagram. (B) Input-output voltage waveforms.

as shown in the shaded areas of the diagram. During overlap time period
t_1, phase A voltage is more positive than phase B voltage, while during
the t_2 interval, phase B is more positive. Diode D_1 will conduct until
time period t_1 ends, then D_2 will conduct beginning at the end of t_1 until
the next area of overlapping is reached.

Note that the voltage across the load device rises to a peak value
twice during each phase alternation of the AC input voltage. These
peaks are 120° apart. Since the DC output voltage never falls to zero,
less AC ripple is present, which results in a purer form of direct current

than single-phase rectifiers produce. The average DC output voltage (V_{DC}) is expressed as

$$V_{DC} = 0.831 \times V_{max}$$

which compares very favorably with single-phase, full-wave rectifier circuits.

A disadvantage of this type of three-phase rectifier circuit is that the AC input lines are not isolated. This lack of *isolation*, which is a direct connection to the AC input lines, could be a hazardous safety factor. To overcome this disadvantage, a transformer can be used, as shown in Figure 3–6, to form a similar three-phase, half-wave rectifier. The secondary voltage can either be increased or decreased by the proper selection of the transformer, providing a variable DC voltage capability. The circuit illustrated used a delta-to-wye-connected transformer. The operation of this circuit is identical to that of the three-phase, half-wave circuit previously discussed; however, line isolation has been accomplished.

THREE-PHASE, FULL-WAVE RECTIFIERS

The full-wave counterpart of the three-phase, half-wave rectifier circuit is shown in Figure 3–7. This type of circuit is popular for many industrial applications. Six rectifiers are required for operation of the circuit. The anodes of D_4, D_5, and D_6 are connected together at point A,

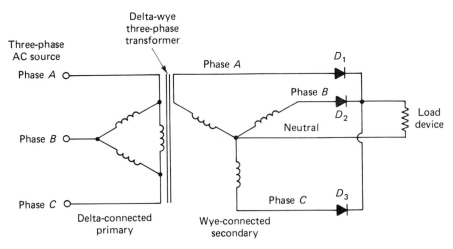

FIGURE 3–6 Three-phase, half-wave rectifier using a transformer.

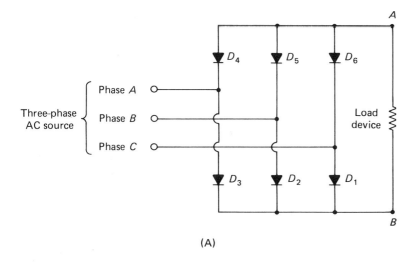

(A)

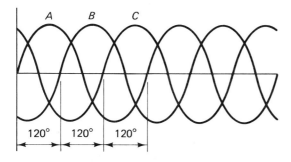

Applied three-phase AC input

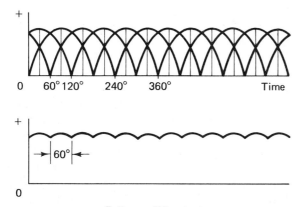

Full-wave DC output

(B)

FIGURE 3–7   Three-phase, full-wave rectifier. (A) Schematic
diagram. (B) Input-output voltage waveforms.

while the cathodes of D_1, D_2, and D_3 are connected together at point B. The load device is connected across these two points. The three-phase AC lines are connected to the anode-cathode junctions of D_1 and D_4, D_2 and D_5, and D_3 and D_6. This circuit does not require the neutral line of the three-phase source; therefore, a delta-connected source could be used.

The resulting DC output voltage of the three-phase, full-wave rectifier circuit is shown in Figure 3–7B. The operation of the circuit is similar to a single-phase bridge rectifier in many respects. At any single instant of time along the three-phase AC input cycle, the anode voltage of one of the diodes is more positive than all others, while the cathode voltage of another diode is more negative than all others. These two diodes will then form the conduction path for that time period. This conducting action is similar to a bridge rectifier, since two diodes conduct during a time interval. Each rectifier in this circuit conducts during one-third of an AC cycle (120°). Peak positive DC output voltage occurs during every 60° of the three-phase AC input.

VACUUM TUBE AND GASEOUS TUBE RECTIFIERS

For many years there have been methods used for the electronic control of equipment. Methods other than semiconductor rectifier circuits are still used to supply DC power. In some cases semiconductors are limited by the amount of current they are capable of handling. Where control of extremely large amounts of current is required, vacuum tube or gaseous tube rectifiers may still be used.

Vacuum tube diodes were used extensively in power supply circuits. A vacuum tube diode, full-wave rectifier circuit is shown in Figure 3–8. Compare this circuit with its semiconductor counterpart of Figure 3–3. A vacuum tube will conduct only when its plate (anode) is made positive relative to its cathode. Tube V_1 will conduct when point A is positive and tube V_2 will conduct when point B becomes positive. A full-wave output will appear across the load device.

A problem of the vacuum tube diode is that not all of the electrons emitted by the cathode reach the plate. The ionization that takes place in a gaseous tube diode overcomes this undesirable aspect of vacuum tube diodes. Thus, an increased amount of current can be conducted in a gaseous tube diode. A gaseous tube rectifier circuit is identical to the vacuum tube circuit shown in Figure 3–8 but with gaseous tube diodes substituted for the vacuum tubes.

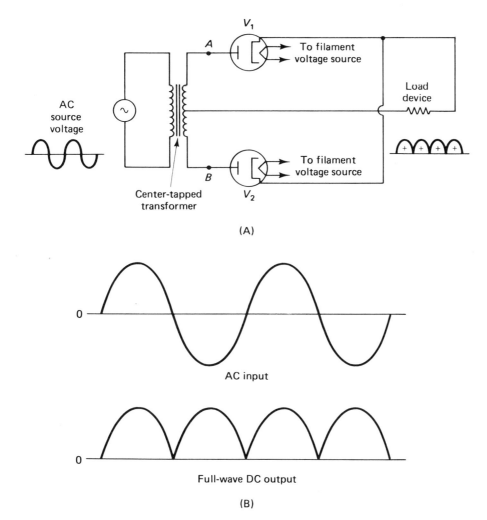

FIGURE 3–8 Vacuum tube diode, full-wave rectifier. (A) Schematic diagram. (B) Input-output voltage waveforms.

POWER SUPPLY FILTER CIRCUITS

The pulsating direct current produced by both single-phase and three-phase rectifier circuits is not pure direct current. A certain amount of AC *ripple* is evident in each type of rectifier. For many applications, a smooth DC output voltage with the AC ripple removed is required. Circuits used to remove AC variations of rectified direct current are called *filter* circuits.

The output of a rectifier has a DC value and an AC ripple value as shown in Figure 3–9. To gain a relative index of the amount of AC variation, the *ripple factor* of the output waveform of a rectifier can be determined by

$$r = \frac{V_{r(rms)}}{V_{DC}}$$

where:

> r is the ripple factor,
>
> $V_{r(rms)}$ is the rms value of the AC component, and
>
> V_{DC} is the average value of the rectified DC voltage.

Another index used to express the amount of AC variation in the output of a rectifier is the *percentage of ripple*:

$$\textbf{Percent ripple} = \frac{V_{r(rms)}}{V_{DC}} \times \textbf{100}$$

A full-wave rectified voltage has less percentage of ripple than does a half-wave rectified voltage. When a DC supply must have a low amount of ripple, a full-wave rectifier circuit is more desirable.

CAPACITOR FILTERS

A simple capacitor filter such as the one shown in Figure 3–10A can be used to smooth the AC ripple of a rectifier output. Figure 3–10C

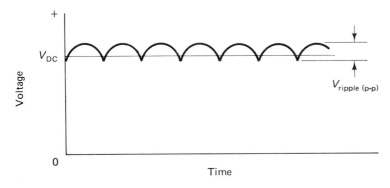

FIGURE 3–9   Rectifier output waveform.

shows the result of adding a capacitor across the output of a 60-Hz, single-phase, full-wave bridge rectifier.

The ideal filtered DC voltage would be one with no AC ripple and a value equal to the peak voltage (V_{max}) of the rectifier output. Note that in Figure 3–10C the value of V_{DC} is approaching that of V_{max}. Compare this to the full-wave-rectified voltage of Figure 3–10B. There are two time intervals shown in Figure 3–10C. Time period t_1 represents diode conduction, which charges the filter capacitor (C) to the peak rectified voltage (V_{max}). Time period t_2 is the time required for the capacitor to discharge through the load (R_L).

A different value of filter capacitor would result with a change in the rate of discharge. If C discharged a very small amount, the value of V_{DC} would be closer to the value of V_{max}. With light loads (high resistance) the capacitor filter will supply a high DC voltage with little ripple. However, with a heavy (low-resistance) load connected, the DC voltage level will drop due to a greater ripple. The increased ripple is caused by the lower-resistance discharge path for the filter capacitor. The effect of a heavier load on the filter capacitor is shown in Figure 3–10D.

By utilizing the values indicated on the waveforms of Figure 3–10, it is possible to express V_{DC} as

$$V_{DC} = V_{max} - \frac{V_{r(p-p)}}{2}$$

The amount of ripple of a 60-Hz, full-wave filter capacitor circuit with a light load is expressed as

$$V_{r(rms)} = \frac{2.4 I_{DC}}{C} \quad \text{or} \quad r = \frac{2.4}{R_L C}$$

where:

I_{DC} is the load current in milliamperes,

C is the filter capacitor value in microfarads, and

R_L is the load resistance in kilohms.

With a heavier load (lower resistance) connected to the filter circuit, more current (I_{DC}) is drawn. As I_{DC} increases, V_{DC} decreases. However, if the value of the filter capacitor (C) is made larger, the value of V_{DC} becomes closer to that of V_{max}. The value of C for a 60-Hz, full-wave rectifier can be determined by using the equation

$$C = \frac{2.4 I_{DC}}{V_{r(rms)}}$$

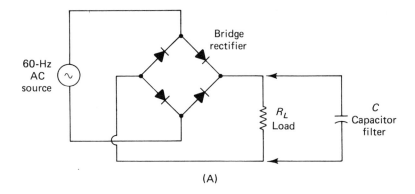

(A)

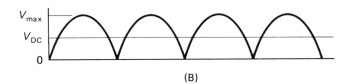

(B)

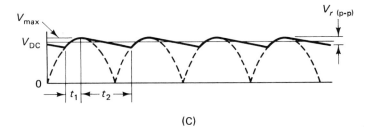

(C)

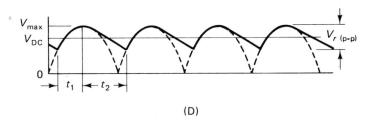

(D)

FIGURE 3–10   Capacitor filter. (A) Bridge rectifier circuit with
filter capacitor. (B) Full-wave output waveform with no filter
capacitor. (C) Full-wave output waveform with filter capacitor
across load. (D) Full-wave output waveform with filter capacitor across heavier (lower-resistance) load.

It should be pointed out, however, that as the value of C increases, the peak value of the current through the diodes will also increase. There is, therefore, a practical limit for the value of the filter capacitor used.

The filter capacitor produces a high DC voltage with low AC ripple for light loads. However, its major disadvantages are higher ripple and lower V_{DC} at heavier loads, poor voltage regulation, and high peak current through the diodes.

RC FILTERS

It is possible to improve upon the previous filter circuit by using an RC filter. Figure 3–11 shows an RC filter circuit. This filter has lower ripple than a capacitor filter does but has a lower average DC voltage due to the voltage dropped across R_1. The purpose of R_1 and C_2 is to add another filter network, which further reduces the ripple. This circuit also operates best with light loads connected. It is possible to have many stages of RC filters to further reduce AC ripple.

PI-TYPE FILTERS

The use of a resistor (R_1) in the RC filter is not desirable in many cases, since it reduces the average DC output of the circuit. To compensate for the DC voltage reduction, a pi-type filter can be used. Figure 3–12 shows this type of filter. The advantage of the choke coil (L_1) over the resistor (R_1) of the RC filter is that it offers only a small DC series resistance, but its AC impedance is much larger. Therefore, it passes direct current and blocks the AC component of the rectified voltage. Several sections of this type of filter can also be used to further reduce AC ripple.

VOLTAGE MULTIPLIER CIRCUITS

The purpose of a voltage multiplier circuit is to increase the value of rectified DC voltage output applied to a load. Voltage multiplier circuits are modifications of filter circuits. The applications of these circuits are lim-

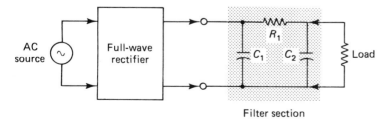

FIGURE 3–11   *RC* filter circuit.

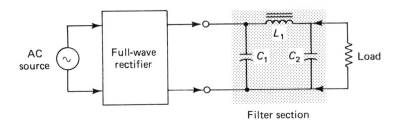

Filter section

FIGURE 3–12   "Pi" filter circuit.

ited; however, there are specific types of equipment that use voltage multiplier circuits.

The basic voltage multiplier circuits are the voltage doublers shown in Figure 3–13. In the *half-wave doubler* circuit shown in Figure 3–13A, when point A has an instantaneous positive charge, diode D_1 will conduct and diode D_2 will be reverse biased. Capacitor C_1 will charge to the peak input voltage (V_{max}). The right side of capacitor C_1 will obtain a positive charge. During the next half-cycle of AC input, the instantaneous polarities will be reversed, placing point B at a positive

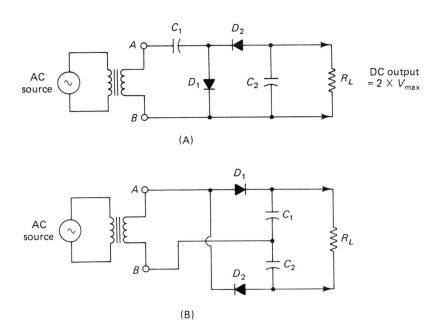

FIGURE 3–13   Voltage doubler circuits. (A) Half-wave doubler. (B) Full-wave doubler.

potential. Diode D_2 will now conduct while D_1 will be reverse biased, placing a charge on capacitor C_2 equal to V_{max}. During the next half-cycle of AC input, the charge on capacitor C_1 will accumulate on capacitor C_2 by discharging through D_2. Also during this half-cycle capacitor C_2 will discharge through the load. The sum of the voltages accumulated on C_2 will be equal to $2V_{max}$. The output is similar to a half-wave rectified waveform that has a simple capacitor filter. The diode PIV rating must be $2V_{max}$ in value.

The operation of the *full-wave voltage doubler* of Figure 3–13B is similar to the half-wave doubler. When point A has a positive potential, diode D_1 will conduct, allowing C_1 to charge to the value of V_{max}. The next half-cycle of AC input will place an instantaneous positive charge on point B. Diode D_2 will then conduct, charging capacitor C_2. With no load connected, the DC voltage at the output terminals is approximately $2V_{max}$. The PIV rating of the diodes must also be $2V_{max}$. More sections of these diode/capacitor circuits can be connected together to form voltage tripler or voltage quadrupler circuits. Extremely high DC voltages can be produced in this way.

POWER SUPPLY REGULATION

The concept of *voltage regulation* can best be understood by referring to the formula

$$\textbf{Percent } VR = \frac{V_{NL} - V_{FL}}{V_{FL}} \times 100$$

where:

VR is the voltage regulation percentage,

V_{NL} is the no-load terminal voltage, and

V_{FL} is the rated full-load terminal voltage.

In an ideally regulated circuit, V_{NL} would equal V_{FL}.

In all types of power supplies that convert alternating current to direct current, the DC output levels are affected by variations in the load. The lower the percentage of regulation (approaching 0 percent), the better regulated the circuit is. For instance, power supplies are capable of having a voltage regulation of less than 0.01 percent, which means that the value of the load has little effect on the DC output voltage produced. A well-regulated DC power supply is necessary for many electronic applications.

ZENER DIODE REGULATORS

A simple voltage regulator circuit uses a zener diode as shown in Figure 3–14. This circuit consists of a series resistor (R_1) and a zener diode (D_1) connected to the output of a rectifier circuit.

Zener diodes are similar to PN junction diodes when they are forward biased. When reverse biased, no conduction takes place until a specific value of reverse breakdown voltage (or "zener" voltage) is reached. The zener diode is designed so that it will operate in the reverse breakdown region of its characteristic curve (see Figure 3–14B). The reverse breakdown voltage is predetermined by the manufacturer. When used as a voltage regulator the zener diode is reverse biased so that it will operate in the breakdown region. In this region changes in current through the diode have little effect on the voltage across it.

The constant-voltage characteristic of a zener diode makes it desirable for use as a regulating device. The circuit of Figure 3–14A is a zener

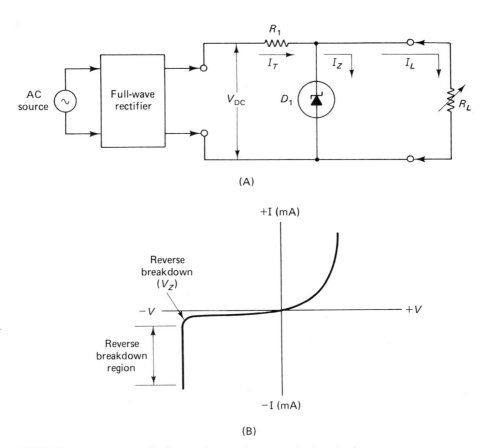

FIGURE 3–14   Zener diode regulator circuit. (A) Circuit diagram. (B) Zener diode characteristic curve.

diode shunt regulator. The zener establishes a constant voltage across the load resistance within a range of rectified DC voltages and output load currents as shown in Figure 3–14B. Over this range, the voltage drop across the zener remains constant. The current flow through the zener (I_z) will vary to compensate for changes in load resistance, since $I_z = I_T - I_L$. Thus, the output voltage will remain constant with variations in load.

TRANSISTOR VOLTAGE REGULATORS

An improvement over the zener diode voltage regulator is a *transistorized series regulator* such as the circuit of Figure 3–15. This regulator has a transistor (Q_1) placed in series with the load device (R_L). Transistor Q_1 acts as a variable resistor to compensate for changes in input voltage. The collector-emitter resistance of Q_1 varies automatically with changes in the circuit conditions. The zener diode establishes the DC bias placed on the base of Q_1. When this circuit is operating properly, as the voltage across the load (R_L) increases, the rise in emitter voltage makes the base less positive. The current through Q_1 will then be reduced, which results in an increase in the collector-emitter resistance of Q_1. The increase in resistance will cause a larger voltage drop across Q_1, which will now compensate for the change in voltage across the load. Opposite conditions would occur if the load voltage were to decrease. Many variations of this circuit are used in regulated power supplies today.

SHUNT TRANSISTOR REGULATORS

Shunt transistor regulators are also used in DC power supplies. The circuit of Figure 3–16 is a shunt voltage regulator. Again, the zener diode (D_1) is used to establish a constant DC bias level. Therefore, volt-

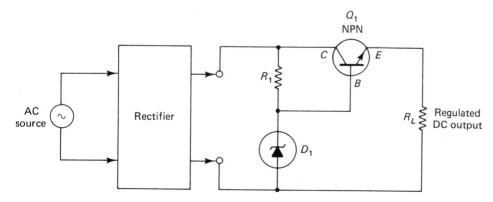

FIGURE 3–15   Series transistor regulator circuit.

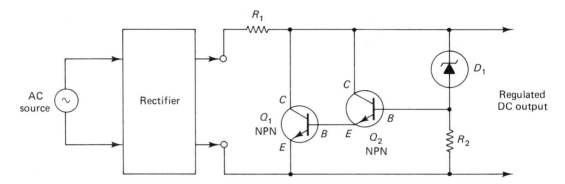

FIGURE 3–16   Shunt transistor regulator circuit.

age variations across the DC output will be sensed only by resistor R_2. If the DC output voltage rises, an increased positive voltage will be present at the base of Q_1. The increased forward bias on Q_1 will cause it to conduct more, which makes the base of Q_2 more positive. Transistor Q_2 will then conduct more heavily. Increased current flow through both transistors causes an increase in the voltage drop across R_1, which will then counterbalance the rise in output voltage. Thus, the DC output voltage will remain stabilized. Decreases in DC output voltage will cause the circuit action to reverse. Shunt transistor voltage regulator circuits of many types are also used extensively in power supplies. The operational characteristics of transistors will be discussed in Chapter 4.

COMPLETE POWER SUPPLY CIRCUITS

A diagram of a complete power supply circuit is shown in Figure 3–17. This power supply has an AC voltage source, a bridge rectifier circuit, a pi-filter circuit, a zener diode regulator, and a load resistance. The complete power supply circuit produces a constant DC voltage output under varying load conditions.

DUAL POWER SUPPLIES

Dual power supplies such as the one shown in Figure 3–18 have both negative and positive outputs with respect to ground. Dual or *split power supplies* were developed as a voltage source for integrated cir-

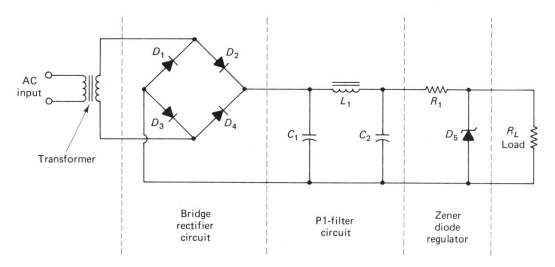

FIGURE 3–17   Complete power supply circuit.

cuits. The secondary winding of the transformer has a center-tap which serves as a common ground connection for the circuit. Each half of the winding supplies a complete full-wave rectifier circuit.

Diodes D_1 and D_2 are rectifiers for the positive DC supply. They are connected to opposite ends of transformer T_1. Diodes D_3 and D_4 are rectifiers for the negative supply and are connected in a reverse direction to the opposite ends of the transformer. The positive and negative DC outputs are with respect to the center-tap of the transformer.

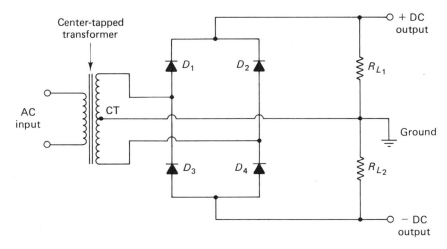

FIGURE 3–18   Dual power supply circuit.

For one alternation assume that the top of the transformer secondary is positive and the bottom negative. Current flows out of the center tap and divides at the junction of R_{L1} and R_{L2}. Equal amounts of current flow in each direction. The top of R_{L1} becomes positive with the bottom of R_{L2} being negative. The path is made complete through D_1 to point A and D_4 to the bottom of the transformer secondary.

The next alternation makes the top of the transformer secondary negative and the bottom positive. Current again flows out of the center-tap and divides at the junction of R_{L1} and R_{L2}. Equal amounts of current again flow in each direction. The top of R_{L1} continues to be positive while the bottom of R_{L2} is negative. One path is made complete through D_3 to the top of the secondary. The other path is through D_2 to the bottom of the secondary winding. The direction of current flow through R_{L1} and R_{L2} is the same as during the first alternation. The DC output voltages of the power supply appear at the top and bottom of R_{L1} and R_{L2}, as shown.

VOLTAGE DIVIDER CIRCUITS

One of the most basic types of circuits used in power supplies is the *voltage divider*. Figure 3–19 shows some types of voltage dividers. The purpose of a voltage divider circuit is to produce specific values of voltage derived from one voltage source. The simple series circuit of Figure 3–19A is a voltage divider. Voltage division takes place due to voltage drops across the three resistors. Since each of the three resistors has the same value (1 kΩ), the voltage drop across each one is 3 V. Thus, a single voltage source is used to derive three separate voltages of a power supply.

Another method used to accomplish voltage division is the *tapped resistor*. This method relies on the use of a resistor which is wire wound and has a tap onto which a wire is attached. The wire is attached so that a certain amount of total resistance of the device appears from the tap to the outer terminals. For example, if the tap is in the center of a 100-Ω wire-wound resistor, the resistance from the tap to either outer terminals is 50 Ω. Tapped resistors often have two or more taps to obtain several combinations of fixed-value resistance. Figure 3–19B shows a tapped resistor used as a voltage divider. In the example, the voltage outputs are each 3 V derived from a 6-V source.

A very common method of voltage division is shown in Figure 3–19C. Potentiometers are used as voltage dividers in volume control circuits of radios and televisions. They may be used to vary voltage from zero to the value of the source voltage. In the example, the voltage output may be varied from 0 to 1.5 V. It is also possible to use a voltage

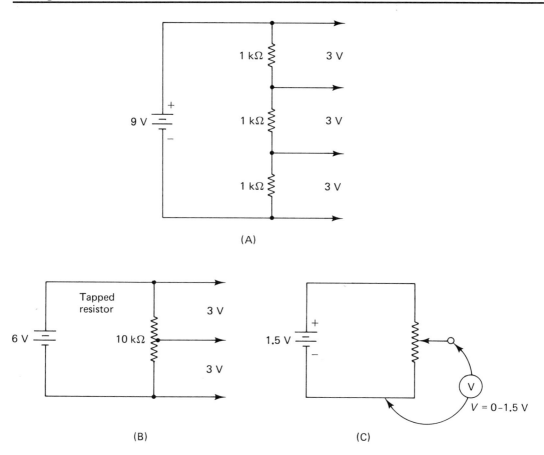

FIGURE 3–19   Voltage divider circuits. (A) Series DC circuit used as a voltage divider. (B) Tapped resistor used as a voltage divider. (C) Potentiometer used as a voltage divider.

divider network and a potentiometer to obtain many variable voltage combinations, as discussed below.

VOLTAGE DIVIDER DESIGN

The design of a voltage divider circuit is a good application of basic electrical theory. Refer to the circuit of Figure 3–20. Resistors R_1, R_2, and R_3 form a voltage divider to provide the proper voltage to three loads. The loads could be transistors of a 9-V portable radio, for example. The operating voltages and currents of the loads are constant. The values of R_1, R_2, and R_3 are calculated to supply proper voltages to each of the loads. The value of current through R_1 is selected as 10 mA. This value

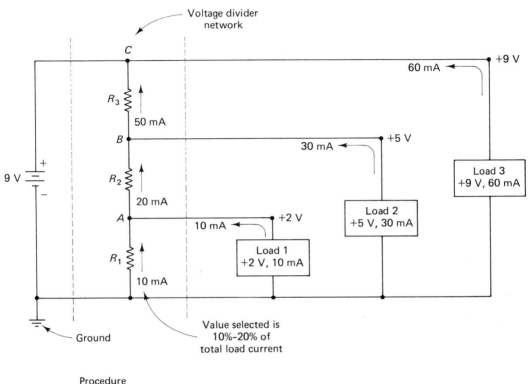

FIGURE 3–20   Voltage divider design.

Procedure

$$R_1 = \frac{E_1}{I_1} = \frac{2 \text{ V}}{10 \text{ mA}} = 200 \ \Omega$$

$$\begin{aligned} P_1 &= E_1 \times I_1 \\ &= 2 \text{ V} \times 10 \text{ mA} \\ &= 20 \text{ mW} = 0.02 \text{ W} \end{aligned}$$

$$R_2 = \frac{E_2}{I_2} = \frac{3 \text{ V}}{20 \text{ mA}} = 150 \ \Omega$$

$$\begin{aligned} P_2 &= E_2 \times I_2 \\ &= 3 \text{ V} \times 20 \text{ mA} \\ &= 60 \text{ mW} = 0.06 \text{ W} \end{aligned}$$

$$R_3 = \frac{E_3}{I_3} = \frac{4 \text{ V}}{50 \text{ mA}} = 80 \ \Omega$$

$$\begin{aligned} P_3 &= E_3 \times I_3 \\ &= 4 \text{ V} \times 50 \text{ mA} \\ &= 200 \text{ mW} = 0.2 \text{ W} \end{aligned}$$

is ordinarily 10–20 percent of the total current flow to the loads (10 + 30 + 60 = 100 mA and 10 percent of 100 mA = 10 mA). The purpose of R_1 is to provide a closed-loop circuit for the voltage divider.

To calculate the values of R_1, R_2, and R_3, the voltage across each resistor and the current through each resistor must be known. Start with R_1 at the bottom of the circuit. The current through R_1 is given (10 mA).

The voltage across R_1 is 2 V since the ground is a zero voltage reference and 2 V must be supplied to load 1. The value of R_1, as shown in the procedure of Figure 3–20, must be 200 Ω (2 V ÷ 10 mA).

Resistor R_2 has a voltage of 3 V across it. Point A has a potential of +2 V and point B has a potential of +5 V for load 2. The *difference in potential* or voltage drop is therefore 5 − 2 = 3 V. The current through R_2 is 20 mA. A current of 10 mA flows up through R_1 and 10 mA flows to point A from load 1. These two currents (10 + 10 = 20 mA) combine and flow through R_2. The value of R_2 must be 150 Ω (3 V/20 mA).

Resistor R_3 has a voltage of 4 V across it (9 − 5 = 4 V). The current through R_3 is 50 mA, since 20 mA flow upward through R_2 and 30 mA flow from load 2 to point B (20 + 30 = 50 mA). The value of R_3 must be 80 Ω (4 V/50 mA).

With the calculated values of R_1, R_2, and R_3 used as a voltage divider network, the proper values of voltage are supplied to the three circuit loads. Minimum power rating of each resistor must also be considered. Minimum power values of 0.02, 0.06, and 0.2 W are calculated in Figure 3–20. Often, a *safety factor* is used to assure that power values are large enough. A safety factor is a multiplier used with the minimum power values. For instance, if a safety factor of 2 is used, the minimum power values for the circuit would become $P_1 = 0.02$ W $\times 2 = 0.04$ W, $P_2 = 0.06$ W $\times 2 = 0.12$ W, $P_3 = 0.2$ W $\times 2 = 0.4$ W.

VOLTAGE DIVISION EQUATION

The voltage division equation, often called the *voltage divider rule,* is convenient to use with voltage divider circuits. The voltage divider equation and a sample problem are shown in Figure 3–21. This equation applies to series circuits. The voltage (E_x) across any resistor in a series circuit is equal to the ratio of that resistance (R_x) to total resistance (R_T) multiplied by the source voltage (E_T).

NEGATIVE VOLTAGE FROM A VOLTAGE DIVIDER CIRCUIT

Voltage divider circuits are often used as power supplies for electronic equipment. In electronics reference is often made to *negative* voltage. The concept of a negative voltage is made clear in Figure 3–22. Voltage is ordinarily measured with respect to a ground reference point. The circuit ground is shown at point A. Often the actual ground reference for electronic circuits is a metal chassis or ground strip on a printed circuit (PC) board to simplify testing and troubleshooting. The negative terminal of a meter can be connected directly to the ground point. All voltage readings are then taken with respect to the ground reference. Point E in Figure 3–22 is connected to the negative side of the power source. Point

$$E_x = \frac{R_x}{R_T} \times E_T$$

E_x = voltage across a resistance

E_T = total voltage applied

R_x = resistance where E_x is measured

R_T = total resistance of voltage divider network

(A)

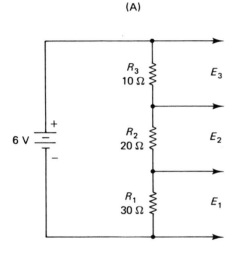

Procedure

$$E_1 = \frac{R_1}{R_T} \times E_T = \frac{30\ \Omega}{60\ \Omega} \times 6\ V = \frac{1}{2} \times 6\ V = 3\ V$$

$$E_2 = \frac{R_2}{R_T} \times E_T = \frac{20\ \Omega}{60\ \Omega} \times 6\ V = \frac{1}{3} \times 6\ V = 2\ V$$

$$E_3 = \frac{R_3}{R_T} \times E_T = \frac{10\ \Omega}{60\ \Omega} \times 6\ V = \frac{1}{6} \times 6\ V = 1\ V$$

(B)

FIGURE 3–21   (A) Voltage division equation and (B) sample problem.

A, where the ground reference is connected, is more positive than point E. Therefore, the voltage across points A and E is − 50 V.

VOLTAGE DIVISION WITH A POTENTIOMETER

A sample problem using a potentiometer is shown in Figure 3–23. A given value of 10 kΩ is used as the potentiometer. The desired variable

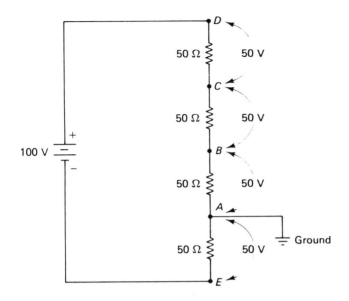

Voltage combinations (taken from ground reference)

E_{AB} = 50 V

E_{AC} = 100 V

E_{AD} = 150 V

E_{AE} = $\boxed{-50 \text{ V}}$

FIGURE 3–22   Negative voltage derived from voltage divider.

voltage from the potentiometer center terminal to ground is 5–10 V. The values of R_1 and R_3 are calculated to derive the desired variable voltage from the potentiometer.

The current flow in a voltage divider network is established by the value of R_2 (10 kΩ) and the range of voltage variation (5–10 V = 5 V variation). The current flow calculation in the circuit of Figure 3–23 is shown in the procedure. Since $I = E/R$, the current through R_2 and the other parts of this series circuit is 0.5 mA. Once the current is found, values of R_1 and R_3 may be found as shown in the procedure. Figure 3–23B shows an easy method of determining voltage drops. A network of resistances in series can be thought of as a scale. In the example the voltage at point A is +5 V and the voltage at point B is +10 V. The difference in potential is 5 V (10 − 5 = 5 V). This is similar to reading a scale.

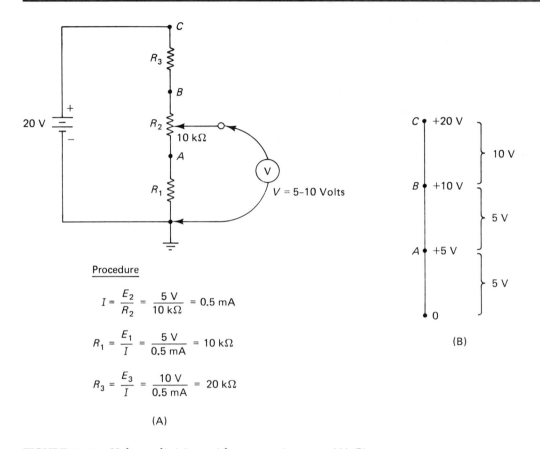

Procedure

$$I = \frac{E_2}{R_2} = \frac{5\text{ V}}{10\text{ k}\Omega} = 0.5\text{ mA}$$

$$R_1 = \frac{E_1}{I} = \frac{5\text{ V}}{0.5\text{ mA}} = 10\text{ k}\Omega$$

$$R_3 = \frac{E_3}{I} = \frac{10\text{ V}}{0.5\text{ mA}} = 20\text{ k}\Omega$$

(A)

FIGURE 3–23   Voltage division with a potentiometer. (A) Circuit. (B) Voltage drops across each resistance considered as a scale.

REVIEW

1.   What methods may be employed to convert alternating current to direct current?

2.   Describe the following types of rectification circuits:

   a) Single-phase, half-wave.

   b) Single-phase, full-wave.

    c) Single-phase bridge.

    d) Three-phase, half-wave.

    e) Three-phase, full-wave.

3. Discuss vacuum tube and gaseous tube rectifiers.

4. What methods may be used to filter direct current?

5. What is a voltage multiplier?

6. What are some methods used for voltage regulation?

7. Discuss the operation of a dual power supply.

8. Explain the functions of each of the circuits used in a complete power supply.

9. What is a voltage divider circuit?

PROBLEMS

1. What is the average value of DC output voltage of a half-wave rectified voltage of 200 V peak value?

2. If a 120-V (rms) voltage is rectified by a half-wave circuit, what is the DC output voltage?

3. What is the DC voltage obtained from a full-wave rectifier if the peak AC input is 150 V?

4. Calculate the ripple factor and percentage of ripple of a filtered rectifier circuit whose DC voltage is 20 V and AC ripple voltage is 1.5 V (rms).

5. The no-load voltage of a DC power supply is 160 V. The full-load voltage is 150 V. What is the voltage regulation?

6. What is the ripple voltage of a full-wave rectifier with a capacitor filter of 25 μF and a load current of 50 mA?

7. A three-phase, half-wave rectifier has an AC voltage input of 240 V (rms). What is the average DC voltage output?

8. Refer to Figure 3–24 and calculate the values of R_1, R_2, and R_3 needed to form the voltage divider.

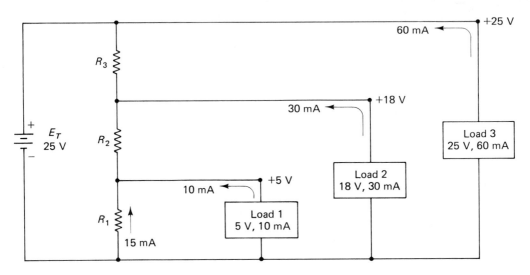

FIGURE 3–24

9. Use the voltage divider values of Figure 3–25. Calculate the minimum power ratings of R_1, R_2, and R_3.

10. Solve for the values of R_1 and R_2 in the voltage divider circuit of Figure 3–25.

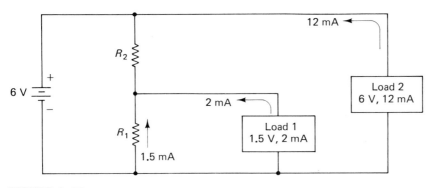

FIGURE 3–25

11. Use the values of Figure 3–25. Calculate the power ratings of R_1 and R_2 with a safety factor of 3 used.

12. Use the voltage division equation to find the values of E_1, E_2, and E_3 in Figure 3–26.

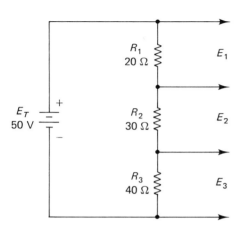

FIGURE 3–26

13. A voltage divider design problem is illustrated in Figure 3–27. Compute the values of R_1 and R_3 needed to obtain a 3- to 12-V variable output.

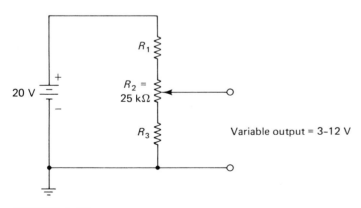

Variable output = 3-12 V

FIGURE 3–27

14. Refer to Figure 3–28. Calculate the values of E_{ab}, E_{bc}, E_{cd}, E_{de}, E_{ac}, E_{ad}, E_{ae}, E_{bd}, E_{be}, and E_{ce}.

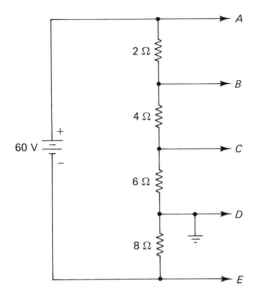

FIGURE 3–28

SUGGESTED LABORATORY ACTIVITIES

The following laboratory activities may be used to supplement the material discussed in Chapter 3.

1. Single-phase, half-wave rectifier.

    a) Construct a single-phase, half-wave rectifier circuit such as the one shown in Figure 3–2. Use values of D_1, R_L, and AC source voltage as specified by the instructor.

    b) Calculate the DC output voltage value.

    c) With a meter measure and record the DC voltages across D_1 and R_L.

    d) With an oscilloscope observe waveforms of the AC source voltage, the voltage across the diode, and the output voltage. Make a sketch of each waveform.

2. Single-phase, full-wave rectifier.

    a) Construct a single-phase, full-wave rectifier circuit such as the one shown in Figure 3–2. Use a center-tapped transformer and values of D_1, D_2, and R_L as specified by the instructor.

b) Calculate the DC output voltage value.

c) With a meter measure and record the DC voltages across D_1, D_2, and R_L.

d) With an oscilloscope observe the waveforms of the AC input, voltage across D_1, voltage across D_2, and output voltage. Make a sketch of each waveform.

3. Single-phase bridge rectifier.

a) Construct a single-phase bridge rectifier such as the one shown in Figure 3–4. Use values of diodes, R_L, and AC input voltage as specified by the instructor.

b) Calculate the DC output voltage value.

c) With a meter measure and record the DC voltages across D_1, D_2, D_3, D_4, and R_L.

d) With an oscilloscope observe the waveforms of the AC input, voltage across each diode, and output voltage. Make a sketch of each waveform.

e) Keep the circuit assembled and proceed to activity 4.

4. Capacitor filter circuit.

a) Add a capacitor of the value specified by the instructor across the output of the bridge rectifier. See Figure 3–10. Be sure to attach the negative side of the capacitor to the negative output of the rectifier. This point is the junction of two anodes of the diodes.

b) With a meter measure and record the DC output voltage with the capacitor connected.

c) With an oscilloscope observe the waveform across R_L. Sketch the waveform.

d) Keep the bridge rectifier circuit assembled.

5. Complete power supply circuit.

a) Add to the bridge rectifier circuit to complete a circuit similar to the power supply shown in Figure 3–17. Use values of C_1, C_2, L_1, D_5, R_1, and R_L as specified by the instructor.

b) With a meter measure and record the DC voltages across C_1, L_1, C_2, R_1, and R_L.

c) Change the value of R_L to a value much larger or smaller than the original R_L.

d) Again, measure and record the DC voltage across R_L.

6. Dual power supply.

a) Construct a circuit similar to the one shown in Figure 3–18. Use a center-tapped transformer and values of diodes, R_{L1}, and R_{L2} as specified by the instructor.

b) With a meter measure and record the DC voltages across the diodes, R_{L1}, and R_{L2}.

c) With an oscilloscope, observe the waveforms across the diodes, R_{L1}, and R_{L2}. Make a sketch of each waveform.

7. Voltage divider circuit.

a) Construct a voltage divider circuit similar to the one shown in Figure 3–19A. Use values of R_1, R_2, and R_3 and a value of source voltage specified by the instructor.

b) With a meter measure E_1, E_2, and E_3. Record the values on a sheet of paper.

c) Calculate the values of E_1, E_2, and E_3 using the voltage division equation.

d) Determine the percent difference between the measured and calculated values of E_1, E_2, and E_3 using the formula

$$\textbf{Percent difference} = \frac{\textbf{calculated value} - \textbf{measured value}}{\textbf{calculated value}} \times \textbf{100}$$

8. Voltage divider design.

a) Construct a voltage divider circuit similar to the one shown in Figure 3–23A. Use values of source voltage, potentiometer R_2, and variable voltage output specified by the instructor.

b) Calculate the values of current, R_1, and R_2 for the circuit. Record these values.

c) Insert values (as close as possible) of R_1 and R_2 into the circuit.

d) With a meter measure the variable voltage output of the circuit to test its operation.

CHAPTER 4
Transistors

The development of the transistor (Brattain and Bardeen, Bell laboratories, in the late forties) did more to cause the advancement of the electronics industry toward miniaturization than any other event or series of events in history. Generally, the transistor served as a replacement for many types of vacuum tubes and was the forerunner in the development of solid-state electronic technology as we know it today. *Solid state* refers to current flow through crystals like germanium or silicon. No vacuum, gas, or heat is involved. The following is an examination of the characteristics of the basic transistor.

PHYSICAL CHARACTERISTICS

The transistor is an electronic device that is constructed of both N- and P-type semiconductor materials, arranged in three distinct locations within the transistor's body. This arrangement of semiconductor materials within the transistor results in it being classified as one of only two transistor types. A transistor may be classified as an NPN transistor or a PNP transistor, Figure 4–1 illustrates each type of transistor, as well as its electrical symbol and lead identification.

As illustrated, a transistor may be composed of two N materials and one P *(NPN transistor)* or two P materials and one N *(PNP transistor)*. In all instances the single material is sandwiched between the other two. Likewise, all NPN or PNP transistors have three external leads. Each lead is affixed to a specific region of the transistor and serves to identify each as the emitter, base, or collector. The base is always the transistor's single and middle semiconductor material.

Transistors may be further classified as *signal* (low-current) or *power* (high-current) transistors and are manufactured in many external shapes known as *outlines*. Some of the most common outlines are illustrated in Figure 4–2. In most instances the shape of the outline and lead placement is used to identify the base, collector, and emitter connections.

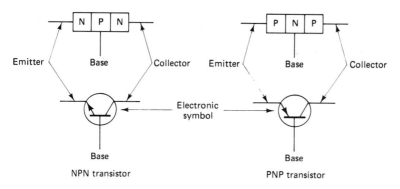

FIGURE 4–1   NPN and PNP transistors.

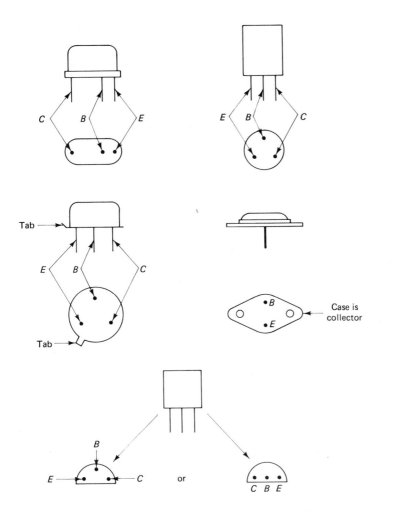

FIGURE 4–2   Transistor outlines (E = emitter; C = collector; B = base).

The transistor is also described as a two-junction device. This can be seen due to the formation of a junction between the emitter-base and base-collector regions of the device. Figure 4–3 illustrates the placement and locations of these junctions.

Just as a depletion zone and potential barrier resulted around the junction of a PN junction diode, likewise the same conditions are found at the junctions of transistors. Each transistor has two depletion zones and two potential barriers. One zone and barrier exists between the emitter and base regions while the second zone and barrier exists between the base and collector regions.

Recall from Chapter 2 that a PN junction diode could be forward biased (cathode negative and anode positive) or reverse biased (cathode positive and anode negative). When forward biased the potential barrier was lowered and the depletion zone was reduced in width resulting in a low forward resistance and an external current flow. When reverse biased the depletion zone was increased in width resulting in a very high reverse resistance and little or no external current.

The depletion zones around the junctions of a transistor react in the same manner as the depletion zone around the junction of the PN junction diode when forward or reverse biased. Figure 4–4 illustrates the forward characteristics of both junctions for both types of transistors. It should be noted that only two of the three external leads are connected at any time. Normally all three leads would be connected into a circuit.

Figure 4–5 illustrates the characteristics of the same junctions when reverse biased. From the illustrations in Figures 4–4 and 4–5 it can be seen that the characteristics of the junctions of both the NPN and PNP transistors are identical when forward or reverse biased. The only difference in the operation of either transistor deals with a reversal of the polarity of biasing voltage. Because of this, the NPN transistor will be used in all future illustrations, discussions, and examples.

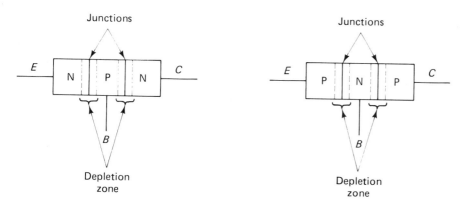

FIGURE 4–3   NPN and PNP junctions.

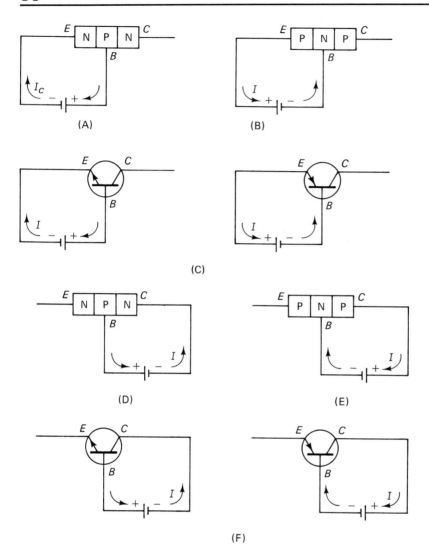

FIGURE 4–4 Junction forward conduction characteristics. (A) Emitter-base junction forward biased; (B) forward resistance low; (C) external current flows. (D) Base-collector junction forward biased; (E) forward resistance low; (F) external current flow. (E = emitter; B = base; C = collector; I = current.)

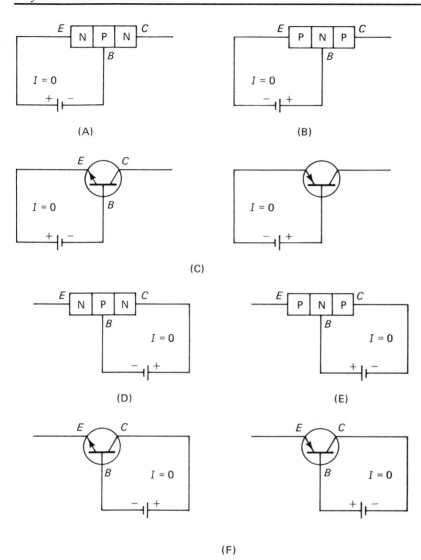

FIGURE 4–5 Junction reverse condution characteristics. (A) Emitter-base junction reverse biased; (B) reverse resistance high; (C) little, if any, external current. (D) Base-collector junction reverse biased; (E) reverse resistance high; (F) little, if any, external current. (E = emitter; B = base; C = collector; I = current.)

ELECTRICAL CHARACTERISTICS

The transistor is a single chip of silicon or germanium that has been doped with an appropriate impurity to form three distinct regions identified as the emitter, base, and collector regions. Each region has both majority carriers and minority carriers. Recall that in the N material the majority carriers are electrons and the minority carriers are holes, while in the P material the majority carriers are holes and the minority carriers are electrons. All three regions, and thus their majority and minority carriers, react and interact when biased correctly. It is important to note at this point that the base region of most transistors is somewhat narrow and lightly doped when compared to the emitter and collector regions. Therefore, there are few majority carriers in this region and even fewer minority carriers.

There are three common methods of connecting a transistor into a circuit. These methods of connection, known as configurations, are the *common-base,* the *common-emitter,* and the *common-collector* configuration. In each configuration the transistor is connected in such a way as to cause one region to be common or interact electrically with the remaining regions.

COMMON-BASE

When a transistor's common-base configuration is used, its base region is common and interacts with both the emitter and collector regions. Figure 4–6 illustrates the common-base configuration. This illus-

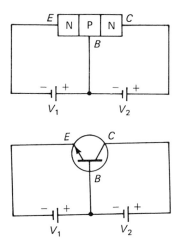

FIGURE 4–6 Transistor common-base configuration. ($E=$ emitter; $C=$ collector; $B=$ base.)

tration shows that in the common-base configuration, the emitter-base junction is forward biased by voltage source 1 (V_1), and the base-collector junction is reverse biased by voltage source 2 (V_2). Recall from Chapters 1 and 2 how majority and minority current carriers react in a forward- or reverse-biased state. Figure 4–7 illustrates this movement of current carriers. When the emitter-base junction is forward biased, the depletion zone around the junction is reduced in width. The reverse-biased state of the base-collector junction results in that depletion zone becoming wider.

The majority carriers (electrons) from the emitter move across the junction and into the base. Since the base is thin and lightly doped, very

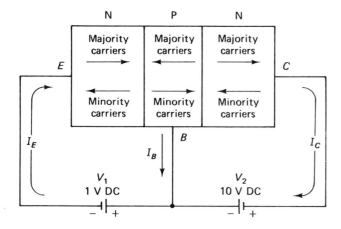

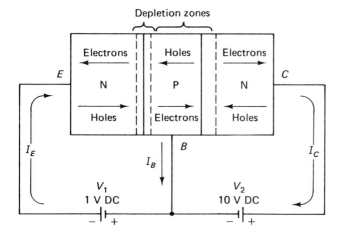

FIGURE 4–7   Transistor current carrier movement. (E=emitter; C=collector; B=base.)

few majority carriers (holes) are available to recombine with these electrons from the emitter. The recombination that does take place appears as a small external base current (I_B). The electrons (majority carriers) from the emitter that do not recombine at the emitter-base junction move into the base as minority carriers and are drawn into the collector by V_2 and ultimately appear as external collector current (I_C). The majority carriers from the collector region do not cause external current. For each electron from the emitter that recombines with a hole at the emitter-base junction or moves into the collector region causing I_C, an electron moves onto the emitter because of V_1. This accounts for the external emitter current (I_E). The minority carriers in all regions are few in number and account for only a very small amount of the external current.

When a transistor is connected in the common-base configuration, its collector current is always slightly less than its emitter current because of the small base current. This is due to the small amount of recombination that takes place at the emitter-base junction. Thus, emitter current minus base current equals collector current ($I_E - I_B = I_C$). Or base current plus collector current equals emitter current ($I_B + I_C = I_E$).

The transistor, when connected in the common-base configuration, demonstrates its *alpha* (α) characteristic, which is an *amplification factor*. Most transistors, with appropriate circuits, are used as amplifiers (to make larger). A transistor connected in the common-base configuration is normally employed as a voltage amplifier. Amplifying devices and amplifier circuits will be discussed in Chapters 7 and 8.

Figure 4–8 illustrates a typical family of collector characteristic curves for a transistor connected in the common-base configuration. As illustrated, the level of emitter current controls the level of collector current. As I_E increases, I_C will increase for any given base-collector voltage. This may be shown by projecting a line from the 6-V point on the base-collector voltage axis until several emitter current (I_E) curves are intersected. From these intersection points a line can be projected to intersect the collector current axis. This indicates that an increased emitter current (I_E) will cause an increased collector current (I_C) for any base-collector voltage. Thus at intersection A, 1 mA of emitter current causes 0.9 mA of collector current. At intersection B, 2 mA of emitter current results in 1.9 mA of collector current.

COMMON-EMITTER

The transistor may also be connected into a circuit with its emitter region being common to its base and collector region. This results in the emitter interacting with both the base and collector of the transistor. Figure 4–9 illustrates the common-emitter configuration. From this illustration we can see that the emitter-base junction is forward biased because of V_1. V_2 provides a strong positive voltage to the collector, making it

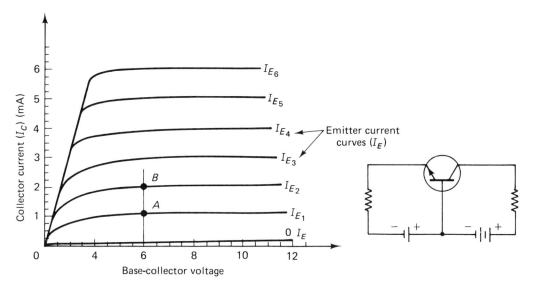

FIGURE 4–8   Typical family of collector characteristic curves.

more positive than the base. Thus, the base would be negative when compared to the collector. Therefore, the base-collector junction is reverse biased.

Even though the transistor is connected differently here than in the common-base configuration, its majority and minority current carriers react the same for the same reasons. Figure 4–10 illustrates the movement of the current carriers when the transistor is connected in the common-emitter configuration. The majority carriers (electrons) move toward the base-emitter junction because of V_1. There they encounter few majority carriers (holes) from the base region. These electrons recombine

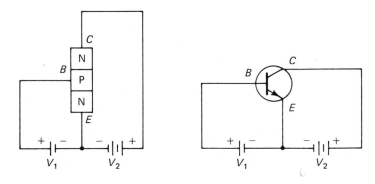

FIGURE 4–9   Common-emitter configuration.

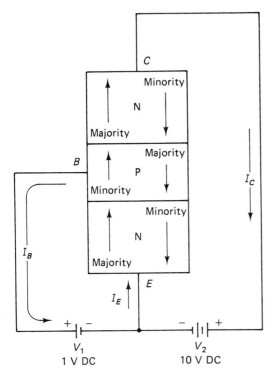

FIGURE 4–10   Common-emitter current carriers. (E = emitter;
B = base; C = collector.)

or are drawn across the base onto the collector side. The positive attraction of V_2 ultimately causes these electrons from the emitter to appear as
external collector current (I_C) flow. Because there is some recombination
at the emitter-base junction, a small external base current (I_B) will flow.
The more electrons that move from the emitter to the base, the larger the
base current (I_B) will be. Likewise the larger the collector current will be.
Thus, the amount of base current (I_B) controls the amount of collector
current (I_C). Figure 4–11 illustrates the common-emitter characteristics
of a typical transistor.

A projected line from the collector-emitter voltage axis intersecting
several base current (I_B) curves will illustrate that an increase in base
current (I_B) will bring about an increase in collector current (I_C) for any
given collector-emitter voltage. Thus, when the collector-emitter voltage
equals 4 V, 0.5 mA of base current will result in 8 mA of collector current (intersection A). At intersection B, 1 mA of base current results in
17 mA of collector current.

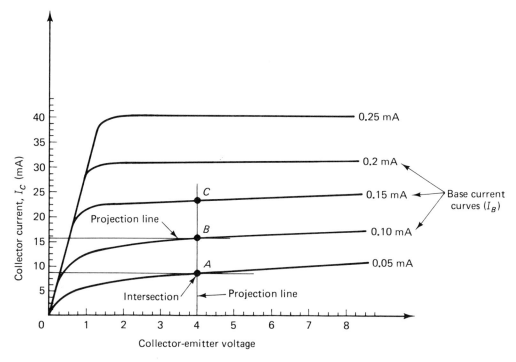

FIGURE 4–11   Transistor common-emitter characteristics.

The transistor, when connected in the common-emitter configuration, demonstrates its *beta* (β) characteristic. The beta characteristic is another amplification factor associated with all transistors. Beta is best used to describe current gain in current amplification.

COMMON-COLLECTOR

When a transistor is connected in such a way as to cause its collector region to be common to both its base and emitter region, the transistor is said to be connected in a common-collector configuration. The common-collector configuration is illustrated in Figure 4–12.

The majority and minority carriers from each region react in the same manner and for the same reasons as previously described. The common-collector characteristics of a transistor are similar to the common-emitter characteristics. When used in amplification circuits, it is sometimes called an *emitter-follower amplifier* and is known for its *impedance-matching* capabilities. Application of transistors connected in the common-collector configuration will be discussed in Chapter 7.

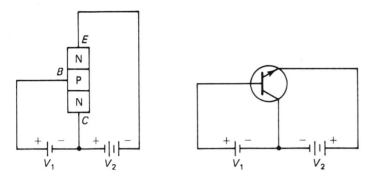

FIGURE 4–12   Transistor common-collector configuration.

MANUFACTURER'S SPECIFICATIONS

A transistor's maximum ratings or operating characteristics may be acquired from the technical data section of an appropriate transistor manual or from a technical specifications sheet supplied by the manufacturer. These specifications include the maximum collector-base voltage, collector-emitter voltage, emitter-base voltage, collector and emitter currents, power dissipation, operating temperature, and base leakage current, among others.

The maximum voltages are the largest voltage potentials that can exist between the regions at any given time. The maximum currents represent the largest currents that can be allowed to pass through the indicated regions. The maximum power dissipation of a transistor is usually indicated at an operating temperature of 25 °C and represents the maximum combinations of currents and voltages associated with its performance. The power dissipation of signal (low-current) transistors is usually listed in milliwatts. Assume the maximum power dissipation of a transistor is listed at 150 mW and it is connected in the common-emitter configuration. When its collector-emitter voltage equals 3 V, its collector current cannot exceed 50 mA.

$$I = \frac{P}{E} = \frac{0.150 \text{ W}}{3 \text{ V}} = 0.05 \text{ A} \quad \text{or} \quad 50 \text{ mA}$$

Likewise when its collector-emitter voltage equals 5 V, its collector current cannot exceed 30 mA.

$$I = \frac{P}{E} = \frac{0.150 \text{ W}}{5 \text{ V}} = 0.03 \text{ A} \quad \text{or} \quad 30 \text{ mA}$$

A transistor's operating temperature is of special interest. Recall that all semiconductors exhibit negative temperature coefficients. That is, as the temperature of its operating environment increases, its resistance will decrease. An operating temperature that exceeds the maximum can cause *thermal runaway* and damage to or destruction of the transistor.

Finally, the *base leakage* (I_{CBO}: collector-emitter current with the base open) characteristic is of interest. This is the maximum current that will flow between the emitter and collector regions, through the base region when the base is not biased. Figure 4–13 illustrates this characteristic.

The base leakage current is small when compared to other maximum current values and is due to the action of the few current carriers in that region.

Figure 4–14 illustrates the absolute maximum characteristics for a typical transistor.

OHMMETER TESTS

A transistor's general condition may be checked by using an ohmmeter. Since an ohmmeter has both a positive and negative lead, it may be affixed to the external leads of a transistor to either forward or reverse bias the transistor's junction, thus indicating forward or reverse resistance. Figures 4–4 and 4–5 illustrate the characteristics of a transistor's junctions when forward or reverse biased. Replacing the battery with an ohmmeter, while observing appropriate polarity, would effectively cause the forward and reverse resistance of the junctions to be measured.

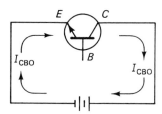

FIGURE 4–13   Transistor base leakage. (E = emitter; B = base; C = collector; I_{CBO} = leakage.)

Collector-base voltage, 25 V

Collector-emitter voltage, 24 V

Emitter-base voltage, 12 V

Collector current, 100 mA

Emitter current, 100 mA

Power dissipation, 150 mW

Operating temperature, 85°C

I_{CBO}, 0.5 μA

FIGURE 4–14   Typical transistor characteristics.

The ohmmeter can also be used to measure base leakage. Figure 4–15 illustrates the ohmmeter connected between the emitter and collector of a transistor. The base remains unconnected for this measurement. In both instances the ohmmeter should indicate a very high resistance because the base leakage is very low. If a transistor exhibits a low resistance when the ohmmeter is connected, as illustrated in Figure 4–15, its condition may be questionable.

In this chapter the transistor's physical and electrical characteristics, common configurations, and manufacturer's specifications have been studied. Answer the following review questions, solve the prob-

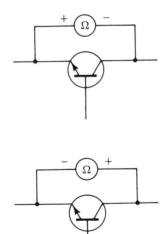

FIGURE 4–15   Ohmmeter checking transistor condition.

lems, and perform the suggested laboratory activities to better understand the characteristics of transistors.

REVIEW

1. What is meant by the term *solid-state* electronics?

2. Why are transistors classified as NPN or PNP devices?

3. What terms are used to identify the three regions of a transistor?

4. Why is a transistor said to be a two-junction device?

5. What is significant about a transistor's outline?

6. What are the differences in the characteristics and operation of an NPN or a PNP transistor?

7. How does the base region of a transistor differ from the emitter and collector regions?

8. What are the characteristics of the common-base configuration?

9. What are the characteristics of the common-emitter configuration?

10. What are the characteristics of the common-collector configuration?

11. How does current flow through the transistor when it is appropriately biased?

12. What is the relation between emitter current and collector current when the transistor is connected in the common-base configuration?

13. What is the relation between base current and collector current when the transistor is connected in the common-emitter configuration?

14. What is the relation among emitter, collector, and base currents when the transistor is connected in the common-base configuration?

15. What configuration is associated with a transistor's alpha characteristic?

16. What configuration is associated with a transistor's beta characteristics?

17. What are manufacturer's maximum transistor ratings and why must they be observed?

18. What is meant by the power dissipation rating of a transistor?

19. What is I_{CBO}?

20. What is the relation between a transistor's operating temperature and thermal runaway?

PROBLEMS

1. Assume a transistor is connected in the common-emitter configuration. Further assume that its maximum power dissipation rating is 200 mW. How much collector current can be permitted when its collector-emitter voltage is 4, 6, 8, and 10 V?

2. Assume a transistor is connected in the common-base configuration. What is its base current when its emitter current is 2 mA and its collector current is 1.92 mA?

3. Assume the transistor in problem 2 exhibited 100 μA of base current at 2.9 mA of collector current. What is its emitter current?

SUGGESTED LABORATORY ACTIVITIES

1. Using an appropriate transistor manual, identify and record the maximum ratings for the following transistors:

   a) 2N408

   b) 2N2896

   c) 2N3773

   d) 2N4036

2. Identify and record the emitter, collector, and base leads of at least three transistors with different outlines.

3. Using the family of characteristic curves illustrated in Figure 4–8, determine when collector currents would result if the base-collector voltage is 8 V and the emitter current is 3, 4, and 5 mA.

4. Using the family of characteristic curves illustrated in Figure 4–11, determine what collector currents would result when the collector-emitter voltage is 6 V and the base current is 0.05, 0.1, 0.15, and 0.2 mA.

5. Using an ohmmeter and a transistor supplied by the instructor, measure and record the forward and reverse resistances of the emitter-base and collector-base junctions.

CHAPTER 5
Triggered Control Devices

Triggered control devices are those electronic components that can be "turned on" or "off" electrically. These devices are generally used to trigger (turn on) circuits that control current flow through a load. In some instances these devices are used to trigger the component that triggers the control circuit.

In this chapter characteristics of the Shockley diode, silicon-controlled rectifier (SCR), diac, triac, and the unijunction transistor (UJT) will be studied. The switching characteristics of the NPN and PNP transistor will also be examined. Finally, the characteristics of the gaseous thyratron and ignitron tubes will be introduced.

SHOCKLEY DIODE

The Shockley diode is a specialized diode that has the characteristics of a switch. It is doped to have two N and two P areas or regions that are arranged alternately. It has two external leads that are identified as the anode and cathode (similar to the PN junction diode) and functions in the forward-biased mode (anode positive and cathode negative). The Shockley diode was named for its inventor, William Shockley, who worked with Bardeen and Brittain in developing the transistor. Figure 5–1 illustrates the physical features and electronic symbol of this specialized device.

It can be seen from the illustration that the Shockley diode exhibits three junctions. As in the PN junction diode and the transistors previously discussed, each junction would have a depletion zone and poten-

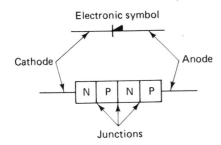

FIGURE 5–1   Shockley diode.

tial barrier. In each region current is due to the action of the majority and minority carriers.

If two PN junction diodes were connected in series with a zener diode, as illustrated in Figure 5–2, it would produce a "make believe" Shockley diode. A better understanding of the Shockley's overall operating characteristics may be gained by reviewing the characteristics of these three diodes.

If the anode of the Shockley is made positive and its cathode negative, D_1 in Figure 5–2 would be forward biased because its cathode would be negative as compared to its anode. Likewise D_2 in Figure 5–2 would be forward biased because its anode would be positive as compared to its cathode. This would cause the depletion zones around junctions 1 and 3 to become narrow. This condition would cause the majority and minority carriers to react accordingly around these junctions and would cause an external current flow if it was not for D_2, connected between D_1 and D_3.

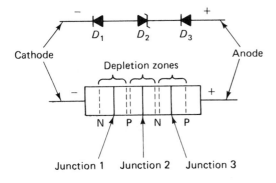

FIGURE 5–2   Internal construction of Shockley diode.

As the positive anode and negative cathode of the Shockley forward biases D_1 and D_3 in Figure 5–2, that same condition reverse biases D_2. This causes the depletion zone around junction 2 to become wider, blocking the general movement of current carriers. The current that might result from the movement of minority carriers would go unnoticed.

As the voltage across the Shockley is increased (anode positive, cathode negative), the breakdown or zener voltage of D_2 is reached. This results in reverse current flow through D_2 and forward current through D_1 and D_3. Thus, the diode is turned on, causing its resistance to drop to a very low level, and the current flow through the diode is limited only by external resistance.

The voltage applied across the Shockley that causes it to be turned on is known as its *switching voltage* and is the same as the reverse breakdown voltage at D_2. Like all zeners, D_2 must have a small *"holding" current* flow in order to keep it on. This holding current is the same as the holding current required to cause the Shockley to remain on. Figure 5–3 illustrates a typical forward characteristic conduction curve for a Shockley diode. The reverse characteristic of the Shockley is similar to that of a PN junction diode.

The most significant ratings or specifications associated with the operation and application of this diode are its switching voltage and its maximum forward current. In some applications its maximum reverse voltage is of interest.

The Shockley diode is used in *oscillator* and *time delay* circuits and can be used to trigger, or turn on, other components such as the SCR. Applications of the Shockley diode will be discussed in Chapter 6.

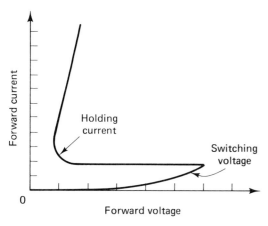

FIGURE 5–3   Forward conduction curve of Shockley diode.

SILICON-CONTROLLED RECTIFIER (SCR)

The SCR is an electronic device used to control the current flow to an external load. Like the Shockley diode it is doped to have two N and two P regions that are arranged alternately. Unlike the Shockley diode it has three external leads identified as the anode, cathode, and gate. The SCR is made of silicon. It is a rectifier whose rectification can be controlled. Thus it is called a silicon-controlled rectifier, or more conveniently an SCR.

Figure 5–4 illustrates the physical characteristics of the SCR. The SCR is always operated in the forward-biased state (anode positive, cathode negative) with a positive voltage of proper value applied to its gate terminal. This positive voltage applied to the gate is used to trigger the SCR as required, thus controlling its conduction. Because its conduction is controlled by its gate potential, it is sometimes classified as a "thyristor" (named after the thyratron tube discussed later in this chapter). Many other solid-state devices having similar characteristics are also classified as thyristors.

It can be seen from Figure 5–4 that the SCR has three junctions. Around each junction there exists a depletion zone. The current carriers around these depletion zones react to a forward or reverse voltage, just as do carriers in most solid-state semiconductor materials previously discussed.

To better analyze how the SCR functions, imagine an NPN and a PNP transistor connected as illustrated in Figure 5–5. From this illustration it can be seen that the base of the PNP transistor (Q_1) and the collector of the NPN transistor (Q_2) are one and the same N material. Like-

Electronic symbol

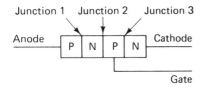

FIGURE 5–4   Construction characteristics of the SCR.

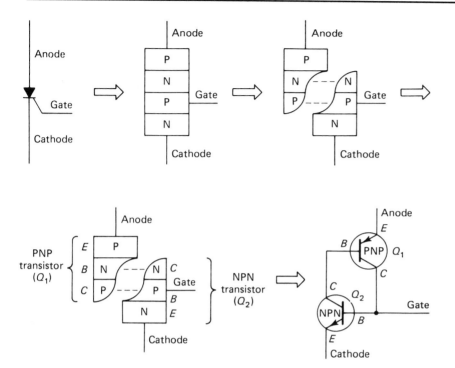

FIGURE 5–5   Analysis of SCR function.

wise it can be seen that the collector of the PNP transistor (Q_1) and the base of the NPN transistor (Q_2) are one and the same P material. As we relate these transistors to the SCR it becomes clear that the emitter of the PNP transistor (Q_1) represents the SCR's anode connection, while the emitter of the NPN transistor (Q_2) represents the SCR's cathode connection. The collector of Q_1 and the base of Q_2 represent the SCR's gate connection.

Figure 5–6 illustrates some very simple circuits that explain the characteristics of the SCR. In figure 5–6A it can be seen that when the SCR's anode is positive and its cathode is negative because of V_1, it is forward biased. Since the gate is not positive as compared to the cathode (gate is not connected), no external current flows in the circuit. This is because within the SCR the base of Q_1 and the base of Q_2 are not forward biased. Recall that current only flows through a transistor when its base is forward biased. Thus, the SCR's forward resistance remains very high. In Figure 5–6B the gate is made positive relative to the cathode by V_2. This forward biases the base of Q_2, which causes emitter-base-collector current within Q_2. The collector current of Q_2 forward biases the base of Q_1, which causes emitter-base-collector current within Q_1. This results

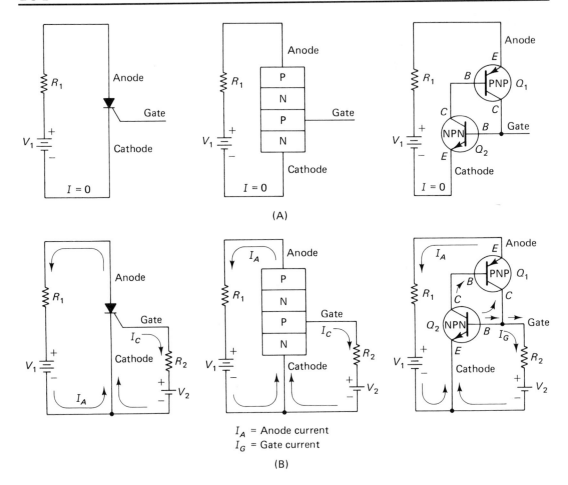

I_A = Anode current
I_G = Gate current

(B)

FIGURE 5–6 Circuit analysis of SCR. (A) Resistance very high. (B) Resistance very low.

in an anode current flow (I_A) onto the cathode and from the anode of the SCR, as well as a gate current flow (I_G) onto the cathode and from the gate of the device.

Generally, the anode current (I_A) is large as compared to the gate current (I_G) and is limited mostly by the value of R_1 since the SCR's resistance drops to a very low level once it is caused to conduct. In normal operation the SCR is always "fired," "triggered," or turned on by causing the gate terminal to be positive as compared to the cathode terminal.

The most remarkable characteristic concerning the SCR is that the gate must be positive for only a very short while in order to cause the SCR to be triggered. Once the SCR has been triggered, the gate potential can be removed and the SCR will remain conductive. Figure 5–7 illustrates this characteristic.

Figure 5–7A illustrates the single-pole, single-throw (SPST) switch in the gate circuit (path between the gate and cathode is closed). This places a positive potential on the gate and causes a gate current (I_G) which triggers the SCR. In Figure 5–7B the SPST switch is open, causing the gate current to cease to flow. However, the anode current (I_A) continues to flow even though the gate has been disconnected. This is true due to the SCR's internal transistors (Q_1 and Q_2 in Figure 5–6) being forward biased, causing virtually all of their current carriers to be set into motion. When the gate potential is removed, these current carriers remain in motion as long as the SCR's anode is positive as compared to its cathode and as long as a minimum amount of current flows from the cathode to the anode. This minimum of anode current (I_A) is known as "holding" current. It is the minimum amount of anode current that will "hold" or keep the SCR turned on. To turn the SCR off once it has been triggered, the positive anode voltage must be removed or the holding current must be allowed to drop below its minimum level. In some instances a very large negative voltage may be applied to the gate of an SCR to turn it off, but this is not commonly done.

Figure 5–8 illustrates a family of forward characteristic conduction curves for a typical SCR. From this family of curves it can be seen that for each anode-cathode voltage, there is a specific value of gate current required to trigger the SCR, as well as a specific holding current required to keep it triggered.

When no gate current exists, a very large anode-cathode voltage (point E) is required to trigger the SCR, and the holding current is at its maximum. With some gate current (I_{G4}) the SCR will trigger with less

I_A = Anode current
I_G = Gate current

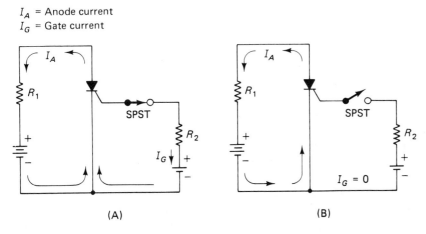

(A)   (B)

FIGURE 5–7   SCR triggered characteristics.

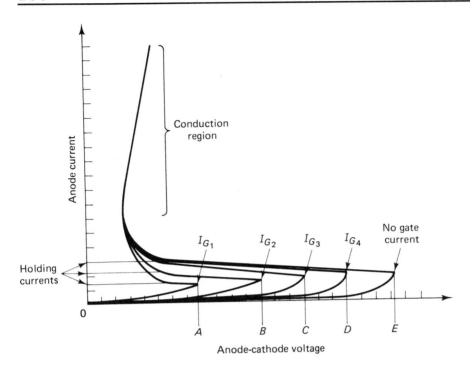

FIGURE 5–8   SCR family of characteristic conduction curves.

anode-cathode voltage (point *D*) and the holding current is reduced. With larger amounts of gate currents (I_{G3}, I_{G2}, and I_{G1}), the SCR will trigger with even less anode-cathode voltages (points *D, C, B,* and *A*) and the holding current is further reduced. Thus, the higher the anode-cathode voltage potential, the less the gate current must be to turn on the SCR. Likewise, the higher the gate current, the less the holding current must be to maintain the SCR's conduction.

OHMMETER TESTS

The general condition of an SCR may be checked with an ohmmeter, as illustrated in Figure 5–9. In Figure 5–9A the resistance of the SCR is generally found to be infinite. This is to say that without a positive gate potential to trigger the SCR on, no current will pass between the anode-cathode of the device.

In Figure 5–9B the SCR's anode-cathode is reverse biased. Anode-cathode current never flows when the SCR anode-cathode is reversed biased; thus its resistance is infinite.

In Figure 5–9C and D the gate and anode of the SCR have alternate polarities of ohmmeter voltage applied. These resistances are either infinite or very high, depending on the SCR being tested. Any indicated

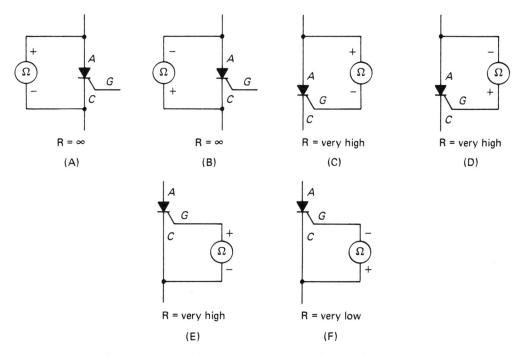

FIGURE 5–9   Ohmmeter tests for SCR. (A) $R=\infty$. (B) $R=\infty$. (C) $R=$ very high. (D) $R=$ very high. (E) $R=$ very low. (F) $R=$ very high.

resistance other than infinite would be due to a very high leakage current between the gate and anode.

In Figure 5–9E the gate-cathode junction is forward biased and is very low. In Figure 5–9F the gate-cathode junction is reverse biased and exhibits a very high or infinite resistance, depending upon the number of minority current carriers found in the affected regions.

Figure 5–10 illustrates a procedure that can be used with most SCRs to test its holding ability. The ohmmeter is connected across the SCR (Figure 5–10A). The positive ohmmeter lead is connected to the SCR's anode while the negative lead is connected to its cathode. A jumper wire is connected between the SCR's gate and its anode (Figure 5–10B). This causes the gate to be positive as compared to the cathode, resulting in the SCR being turned on. When turned on the SCR's anode-cathode resistance drops to a very low level.

In Figure 5–10C the jumper wire has been removed, yet the SCR's anode-cathode resistance is still very low. This indicates that the SCR has been triggered and remains conductive as long as its anode is positive, its cathode is negative, and the minimum holding current is maintained. Thus it holds in the "on" position even though the positive gate potential has been removed.

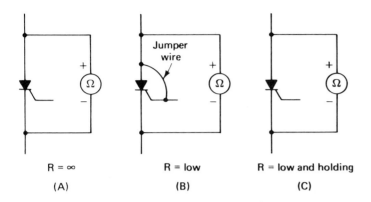

R = ∞   R = low   R = low and holding

(A)   (B)   (C)

FIGURE 5–10   Ohmmeter checking SCR's "holding" ability.
(A) $R=\infty$. (B) $R=$low. (C) $R=$low and holding.

An SCR that does not exhibit the general resistance characteristics as previously explained may be of questionable condition. These resistance characteristics may vary between low- and high-current SCRs.

MANUFACTURER'S SPECIFICATIONS

There are several manufacturer's specifications that are important to know when selecting an SCR to function in a specific circuit. Some of the most important are identified as the forward breakover voltage, forward current, gate turn-on (trigger) current, and holding current. Generally, these ratings are listed by the manufacturer in reference to a specific operating temperature.

The forward breakover voltage is the anode-cathode voltage (anode positive, cathode negative) that will cause the SCR to be turned on in the absence of a gate current. The forward current is the maximum current that can be conducted between the SCR's anode-cathode once it has been triggered. The gate turn-on current is the amount of gate current required to cause the SCR to be triggered when the anode-cathode voltage is at a specific value. The holding current is the minimum amount of current that must pass between the cathode-anode of the SCR in order to cause it to remain in the on, or conductive, state.

DIAC

The diac is a solid-state electronic device that acts as a switch when appropriately biased in either direction. Because it can be triggered to conduct in either direction, it is more generally referred to as an AC switch (conducts in either direction) rather than a DC switch (conducts in only one direction). Figure 5–11 illustrates the characteristics of the diac.

Figure 5–11A illustrates one of the widely accepted electronic symbols of the diac. Figure 5–11B illustrates the diac to be a three-region (NPN) device, having two identical junctions and two external leads identified as anode 1 and anode 2.

Figure 5–12 indicates the diac to be similar to two zener diodes connected in series, back-to-back. If these are identified as D_1 and D_2 and alternate voltage polarities are applied to anodes 1 and 2, a better understanding of how the diac functions can be gained.

Recall that the zener diode will readily conduct current when forward biased (anode positive, cathode negative). Likewise, recall that a zener diode will also conduct current when reverse biased, assuming its reverse breakdown voltage or zener voltage is equaled or exceeded. Figure 5–12B illustrates that anode 1 is positive with respect to anode 2. This forward biases D_2 and reverse biases D_1. If this voltage exceeds the reverse breakdown voltage or zener voltage of D_1 (sometimes 30 V or more), it will break into conduction. Since D_2 is forward biased by virtue of the same voltage potential, it too will conduct. Thus, when the voltage potential between anode 1 and anode 2 exceeds the zener voltage of D_1 in our illustration, the diac's resistance will drop to a low level and the diac will readily conduct current.

Figure 5–12C illustrates the action of the diac when anode 1 is negative with respect to anode 2. This results in D_1 being forward biased while D_2 is reverse biased. When the voltage between anodes 1 and 2 exceeds the reverse breakdown or zener voltage of D_2 (again sometimes 30 V or so), the diac's resistance is reduced to a very low level and the device conducts in the opposite direction. Since both junctions of the diac are formed due to the identical doping of its regions, identical volt-

Electronic symbol

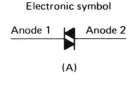

(A)

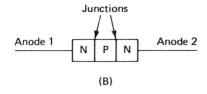

(B)

FIGURE 5–11   Diac construction characteristics.

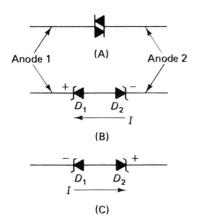

FIGURE 5–12   Diac/zener functions.

age potentials across its anodes of either polarity will result in the diac conducting current in either direction once it has been turned on.

The simple circuit illustrated in Figure 5–13 shows the diac's characteristics when an appropriate peak-to-peak AC voltage is applied to its anodes.

Figure 5–14 illustrates a typical voltage-current characteristic curve of a diac. As shown in the illustration, the diac will readily conduct in both directions once its breakover voltage (forward or reverse) has been reached or exceeded.

MANUFACTURER'S SPECIFICATIONS

Manufacturer's specifications for the diac are its electrical characteristics and values indicated at a specific operating temperature (usually 25 °C). The most important specifications include the breakover voltage (forward or reverse), peak current, peak breakover current, and off-state current at a specific test voltage.

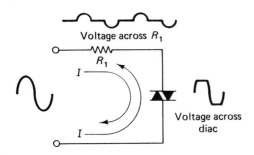

FIGURE 5–13   Circuit analysis of diac.

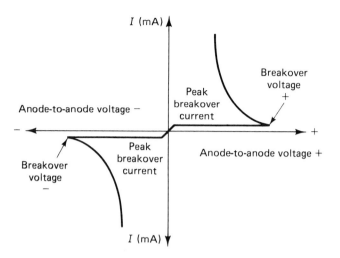

FIGURE 5–14   Diac voltage-current characteristics.

The *forward or reverse breakover* voltage is the voltage between anodes 1 and 2 that will cause the diac to conduct. The *peak current* is the maximum current that can be allowed to pass through the diac during a specific pulse rate duration (length of time). The *peak breakover current* is the maximum current flow through the diac at its breakover voltage. The *off-state current* is the amount of leakage current flow through the diac before it breaks into conduction and is usually listed in relation to a specific test voltage. Figure 5–15 lists typical manufacturer's specifications of a typical diac.

The diac is used to trigger SCRs or triacs. Applications of the diac will be explored in Chapter 6.

TRIAC

The triac is a solid-state electronic device that is used to control AC load current. It has three terminals identified as anode 1, anode 2, and gate and functions somewhat like two identical SCRs connected back-to-back

Breakover voltage	29–35 V (typical, 32 V)
Peak current	2 A (30 μs pulse duration)
Peak breakover current	50 μA (at breakover voltage)
Off-state current	10 μA (at 20 V)

FIGURE 5–15   Manufacture specifications of diac.

and sharing a common gate. Figure 5–16 illustrates the characteristics of the triac.

The unique characteristic that causes the triac to react differently from SCRs is that either a positive or negative gate potential will cause the triac to trigger and conduct current in either direction. This is primarily due to the internal arrangement of the shared semiconductor regions and the doping of these regions. Thus, any polarity of anode-to-anode voltage and any polarity of gate voltage would result in conduction, as illustrated in Figure 5–17.

An AC voltage applied across the anodes of the triac, as well as to its gate terminal, would cause conduction, as illustrated in Figure 5–18. Carefully controlling the time at which the gate potential is applied to the triac would control the length of time the triac would conduct during any AC cycle. This would effectively control the AC current to a series load and can be accomplished by a number of methods. These methods of control will be discussed in Chapter 6.

Figure 5–19 illustrates the characteristic conduction curve of a typical triac. Like the SCR and other devices, a minimum holding current must be maintained in either direction in order to keep the triac in the "on" state.

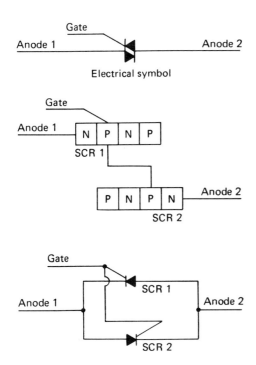

FIGURE 5–16   Triac construction characteristics.

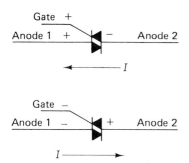

FIGURE 5–17   Triggered characteristics of triac. (I = anode-to-anode current.)

MANUFACTURER'S SPECIFICATIONS

Manufacturer's specifications for the triac are generally listed in relation to a specific operating temperature and a specific operating frequency. Generally, the operating temperature associated with the specifications is 25 °C, although this may vary due to the triac being able to conduct large currents. Most operating frequencies are listed as 50–60 Hz, although some triacs are designed to function quite well at 400 Hz.

Since the triac is primarily an AC-operated device, its specifications are usually listed in instantaneous, or rms, values. Specifications of interest include repetitive peak off-state voltage, rms on-state current, peak surge on-state current, peak gate-trigger current, gate power, holding current, and operating temperature.

The *repetitive peak off-state voltage* is the maximum instantaneous voltage that may be applied between anodes which will not turn the triac on when the gate is open. The *rms on-state current* is the maximum amount of current that can pass from anode to anode when the triac is conducting. The *peak surge on-state current* is the maximum instanta-

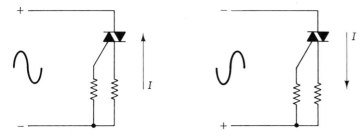

FIGURE 5–18   AC applied with resulting triac conduction. (I = anode-to-anode current.)

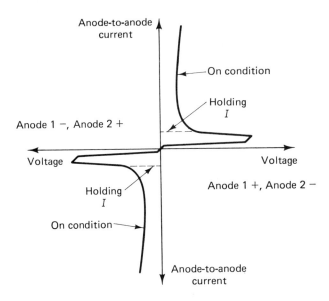

FIGURE 5–19   Triac conduction curve.

neous surge current a triac can withstand without being damaged. The *peak gate-trigger current* is the maximum instantaneous gate current under specified conditions. The *gate power rating* is the maximum instantaneous peak power dissipation of the triac's gate within a specified time, usually listed in watts per microsecond. Finally the *holding current* is the minimum current required to cause the triac to remain in the on state when the gate is open. The operating temperature may range to over 100 °C and is critical since this device, like most other solid-state semiconductor devices, exhibits a negative temperature coefficient.

UNIJUNCTION TRANSISTOR (UJT)

The unijunction transistor (UJT) is a specialized solid-state device that has only two regions (P- and N-type materials), and three external leads identified as base 1, base 2, and emitter. Even though this device is called a transistor, it is not used in amplification applications. It is used as a triggered device to be turned on or to turn on other devices such as the SCR. Figure 5–20 illustrates the characteristics of the UJT. Since *uni* means one and this device has only one junction, it is a one-junction component or unijunction device. The term *transistor* is technically in-

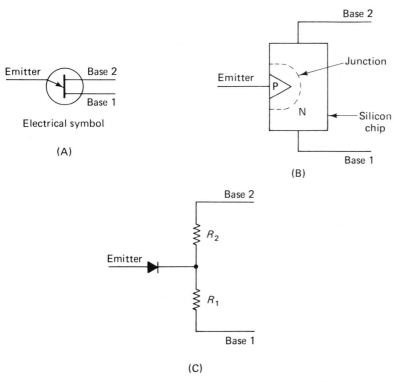

FIGURE 5–20   Construction characteristics of UJT.

correct but will be used throughout this section because of its wide acceptance.

From Figure 5–20B it can be seen that the basic UJT is no more than an N-type silicon chip with a small area of P-type material doped at a specific location on the chip. Because of its construction a junction is formed between the P and N materials. Current flow is due to the movement of majority and minority current carriers as in other solid-state devices.

Figure 5–20C best illustrates the internal characteristics of the UJT, using individual components. The combined resistances of R_1 and R_2 represent the total resistance of the N material between bases 1 and 2. R_1 represents the resistance between base 1 and the point on the silicon chip where the emitter (P material) is located. R_2 represents the resistance between base 2 and the location of the emitter. The illustrated diode represents the junction formed by the emitter and the silicon chip.

The ratio between the value of R_1 to the total resistance is an important characteristic of all UJTs and is known as the *intrinsic stand-off ratio* (η). Thus, if R_1 is 6000 Ω and R_2 is 4000 Ω, then the total resistance between bases 1 and 2 is 10,000 Ω (6000 + 4000 = 10,000 Ω). Since R_1 is 6000 Ω and total resistance is 10,000 Ω, the intrinsic stand-off ratio (η) equals 0.6 ($\eta = R_1/R_T = 6000/10,000 = 0.6$ Ω). The intrinsic stand-off ratio of a UJT determines the current flow between the emitter and base 1.

The combined resistances of R_1 and R_2 act as a voltage divider, as illustrated in Figure 5–21. During the normal operation of the UJT, base 2 is positive with respect to base 1 and a voltage drop develops across each. The sum of the voltage drops equals the source voltage.

Applying Ohm's law to the circuit illustrated in Figure 5–21 results in a total current equal to 1 mA ($I = E/R = 10$ V/10,000 $\Omega = 0.001$ A, or 1 mA). The voltage drop across R_1 equals 6 V ($E = I \times R = 0.001$ A $\times$ 6000 $\Omega = 6$ V). The voltage drop across R_2 equals 4 V ($E = I \times R = 0.001$ A $\times$ 4000 $\Omega = 4$ V). The sum of the voltages across R_1 and R_2 equals the source voltage ($E_{R1} + E_{R2} = E_S$, or 6 + 4 = 10 V).

In order to cause conduction between the emitter junction and base 1, a voltage greater than the voltage drop developed across base 1 must be connected between the emitter and base 1, as illustrated in Figure 5–22. This is true because the emitter junction is reverse biased by the voltage across base 1, which is +6 V. This voltage drop and the intrinsic stand-off voltage are one and the same and may be determined by multiplying η (intrinsic stand-off ratio) by the V_1 source voltage (0.6 $\times$ 10 V = 6 V).

Thus, to cause current flow between base 1 and the emitter, voltage source V_2 must be greater than 6 V because of the few tenths of a volt required to overcome the potential barrier of the junction formed by the emitter and the silicon chip.

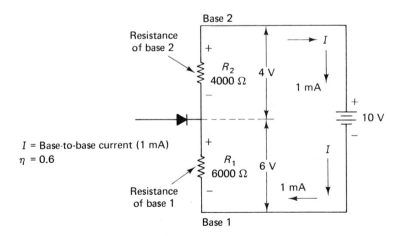

FIGURE 5–21   Voltage divider characteristics of UJT.

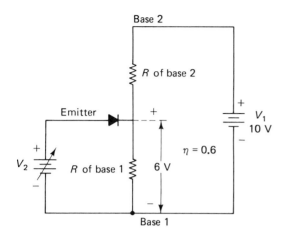

FIGURE 5–22   Intrinsic standoff ratio voltage characteristics.

Consider the circuit illustrated in Figure 5–23. The intrinsic stand-off ratio of the UJT is 0.7. V_1 represents the emitter–base 1 voltage. R_1 and R_2 are external resistors inserted to control current.

Figure 5–23A illustrates S_1 closed and S_2 open. This would cause a base-to-base current (I_1) limited by the resistance of R_1, base-to-base resistance, and R_2. A voltage drop would be developed across R_1, the base-to-base resistance of the UJT, and R_2. The sum of these voltages would equal the source voltage V_1. The base-to-base voltage of the UJT multiplied by η and then added to the voltage across R_1 would equal the V_x voltage required to forward bias the emitter junction and cause emitter–base 1 current. This ignores the few tenths of a volt required to overcome the potential barrier of the emitter junction.

Figure 5–23B illustrates S_1 and S_2 closed. Assuming V_2 is slightly greater than V_x, the emitter–base 1 junction would be forward biased. This would result in a sharp decrease in emitter–base 1 resistance and in emitter–base 1 current (I_2). The emitter–base 1 current (I_2) would flow through R_1, sharply increasing the voltage drop across R_1. With S_2 opened again, emitter–base 1 current would cease to flow and the voltage across R_1 would return to its original level.

Figure 5–24 illustrates an emitter characteristic conduction curve for a typical UJT. A portion of this curve indicates that at the point the emitter becomes forward biased, the emitter–base 1 resistance sharply decreases even though the emitter voltage is increasing. This device exhibits the same negative resistance characteristic of many other solid-state devices once it has been triggered.

The UJT is used in oscillation circuits and as a trigger device for turning on other components such as the SCR. Applications of the UJT will be examined in Chapter 6.

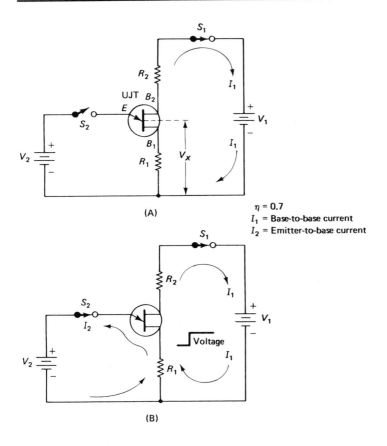

$\eta = 0.7$
I_1 = Base-to-base current
I_2 = Emitter-to-base current

FIGURE 5–23   Typical UJT DC conduction characteristics.

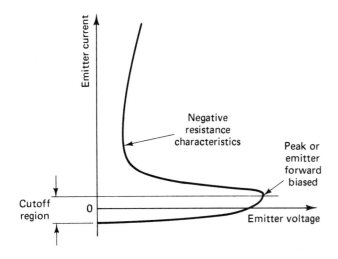

FIGURE 5–24   UJT emitter characteristics conduction curve.

MANUFACTURER'S SPECIFICATIONS

The more important manufacturer's specifications concerning the UJT are the maximum emitter reverse voltage, emitter current, interbase voltage, and intrinsic stand-off ratio. Like other solid-state devices these values are listed with the UJT's operating temperature established at 25 °C. The characteristics of the UJT may be altered significantly with a change in operating temperature.

NPN-PNP TRANSISTORS

Recall from Chapter 4 the characteristics of the NPN-PNP transistors. When the transistor is considered as a triggered control device, it is important to examine the two extreme states associated with the transistor's operation. These two states are the transistor's saturation and cutoff states.

Consider the family of characteristic curves illustrated in Figure 5–25. These curves represent the characteristics of a transistor connected in the common-emitter configuration. As indicated in previous chapters, an increase in base current will result in an increase in collector current.

From the characteristic curves illustrated in Figure 5–25, it can be

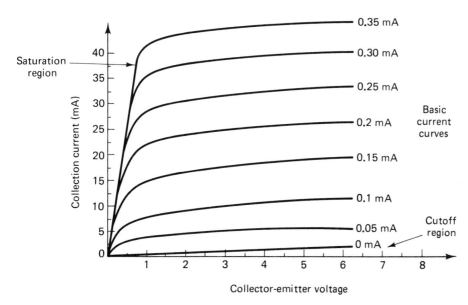

FIGURE 5–25   Family of common-emitter transistor characteristic curves.

seen that when no base current flows, no collector current will flow. For all practical purposes, the transistor will be nonconductive or will be cut off. This coincides with the cutoff region of the characteristic curves.

These same characteristic curves indicate that with a proportionally large base current, the transistor is turned on and the collector current rises to a rather significant level. This coincides with the saturation region of the transistor's characteristics. The transistor may or may not saturate in this region, but its conduction level will be significantly increased from zero to many milliamperes. Thus, the transistor may be turned on or off at a very rapid rate (at the speed of light).

Figure 5–26 illustrates a circuit that shows the DC switching characteristics of a transistor. In the illustrated circuit it can be seen that when the input equals zero, no base current, and thus no collector current, flows and the collector-emitter (CE) voltage equals V_1 or the source voltage. With no collector current, there is no voltage drop across R_1, thus CE voltage equals V_1 voltage. With an input, the base is strongly forward biased (base positive, emitter negative) which causes a larger base current flow. The base current causes the collector current to rise to a high level. This sudden increase in collector current results in an

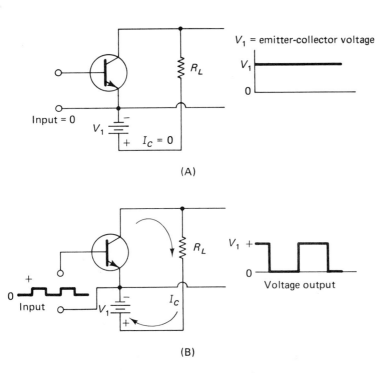

FIGURE 5–26 Transistor DC switching characteristics. ($I_C =$ collector current.)

increased voltage drop across R_1 and a decreased CE voltage across the transistor. Thus, a series of outputs (variations in CE voltage) would result as long as the input to the base was maintained. The transistor would be constantly changed from an "on" to an "off" state.

The voltage variation that results from a transistor being turned on and off can be used to trigger other components and/or circuits. Before the development of integrated circuit devices, the triggering or switching characteristics of the transistor were widely used in logic circuitry. They are still being used in computers and computer electronics today.

Applications concerning the transistor being used as a trigger device and switch will be explored in Chapter 6.

THYRATRON

The thyratron is a gaseous tube that is classified as a controlled rectifier and is very similar to the SCR as far as its characteristics are concerned. Because the two devices are similar in function, the SCR has largely replaced the thyratron tube in most electronic circuitry.

Even though the SCR is considered to be the state of the art as far as controlled rectifiers are concerned, the thyratron tube is still being used in the manufacturing and communications industries in certain specialized applications. For this reason, its characteristics will be explained.

The thyratron tube is a gaseous tube that functions similarly to the hot-cathode gaseous diode tube studied in Chapter 2 with one exception. The thyratron tube has an additional element known as the control grid, as illustrated in Figure 5–27.

Recall how gas ionizes within a closed chamber. When the filament of the thyratron has a voltage applied that results in current flow, the *filament* heats up. This heat is transferred to the cathode which emits

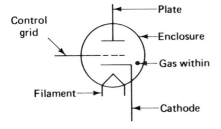

FIGURE 5–27   Elements of thyratron.

electrons. The electrons form a *space cloud* around the cathode. When the plate is made positive with respect to the cathode, the electrons from the space cloud would normally be attracted to and move toward the positive plate. This is not true with the thyratron because of the effect of the *grid* upon this movement of electrons.

The grid is constructed of fine mesh wire that is placed between the cathode and plate of the thyratron. Since the grid is exactly in the path of electron movement from the cathode to the plate, it can control this movement if it is negative with respect to the cathode. This negative potential placed upon the thyratron's grid repels electrons back toward the cathode's space cloud and for all practical purposes does not allow ionization to take place.

As the plate becomes more positive, or the grid becomes less negative, a few electrons are attracted through the grid. Because of bombardment and collision, additional electrons are freed and are likewise attracted to the plate. This effect is cumulative and results in the ionization of the gas within the tube's enclosure. When ionization takes place, most of the electrons are attracted to the plate and appear as external current limited only by the external load. The positive ions that result due to ionization are attracted to the space cloud around the cathode and to the control grid. These positive ions form a sheath around the negative grid causing it to lose control over the tube's current once ionization has taken place. Thus, it becomes obvious that the negative grid of the thyratron can only delay the "firing," or ionization, of the tube. After ionization the grid loses all control. To turn the tube off, one must remove the anode voltage, which results in deionization of the tube.

Figure 5–28 illustrates the plate-grid voltage characteristic curve of the thyratron. From the curve illustrated in Figure 5–28 it becomes obvious that, for any value of negative grid voltage, there is a specific positive anode voltage that will cause the tube to ionize. The more negative the grid, the more positive the plate must be in order to cause the tube to ionize. At point A on the curve, when the grid is -4 V, the plate must be $+40$ V to cause ionization. Likewise points B, C, and D indicate that when the grid is -6, -8, and -10 V, respectively, the plate voltage must equal $+60$, $+80$, and $+100$ V (or more) to cause ionization. If the grid voltage is 0, the tube would ionize at a very low level of plate voltage. The ionization level of the tube without grid voltage would be determined by the tube's physical characteristics and the gas within the tube's enclosure.

Figure 5–29 illustrates characteristic conduction curves of a typical thyratron tube. The characteristics indicate that the action of the thyratron tube is very similar to that of the gaseous hot-cathode diode except for the influence of the grid. Once the tube ionizes, its resistance de-

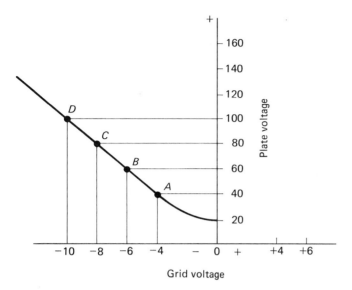

FIGURE 5–28   Thyratron grid voltage characteristics.

creases and the voltage across the tube drops to a very low level. The voltage across the tube remains constant until the tube deionizes.

Like the hot-cathode gaseous diode tube, the thyratron must be allowed to warm up before the plate voltage is applied. Likewise it must never be operated without an external load large enough to limit the tube's current.

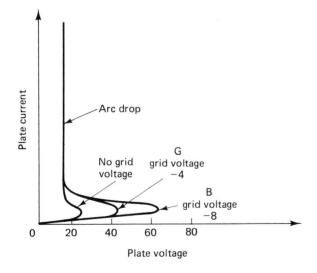

FIGURE 5–29   Thyratron characteristic conduction curves.

Like the SCR, the thyratron tube is used to control a pulsating DC current to a load. The thyratron's application will be examined in Chapter 6.

IGNITRON

The ignitron is a gaseous tube that is designed to control and conduct several thousands of amperes of current and is used in industries where large amounts of electric energy are converted to thermal energy. The ignitron contains three elements identified as the plate, cathode, and ignitor, as illustrated in Figure 5–30.

The unique characteristic associated with this tube is that it contains no filament. The cathode is actually a pool of liquid mercury, and the *ignitor* is used to create an arc between the mercury pool and its tip to cause ionization.

As in all gaseous rectifiers, current only flows from the cathode to the plate once ionization takes place. To cause ionization, the ignitor is made positive relative to the negative cathode (mercury pool). When the ignitor becomes positive enough, an electric arc is created between the ignitor and the cathode. This results in current flow to the ignitor, which ionizes the gas within the tube. Once the tube ionizes, the resistance between the negative cathode and positive plate becomes much less than the resistance of the ignitor circuit. This allows the major portion of current from the cathode to pass directly to the plate. Thus, by controlling the positive ignitor potential, the ionization and conduction of the tube are controlled. This controls the amount of current flowing within the tube and to its external load. Like the other gaseous tubes, this device should never be operated without an external load.

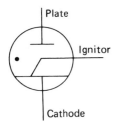

FIGURE 5–30   Ignitron elements.

In this chapter the characteristics of the Shockley diode, diac, triac, SCR, UJT, transistors as triggering devices, and the thyratron and ignitron tubes have been examined. Answer the following review questions, solve the problems, and perform the suggested laboratory activities to help better understand the characteristics of the devices discussed in this chapter.

REVIEW

1. Why is the Shockley diode said to have the characteristics of a single-pole, single-throw switch?

2. How many solid-state semiconductor regions and junctions are associated with the Shockley diode?

3. What does the term *holding current* mean in reference to the operation of the Shockley diode?

4. What are the most significant manufacturer's specifications associated with the operation of the Shockley diode?

5. Where does the SCR get its name?

6. How do the physical characteristics of SCR differ from those of the Shockley diode?

7. What role does the gate of the SCR play in triggering, or turning on, this device?

8. What is a *thyristor*?

9. Why is an SCR like an NPN and PNP transistor connected with shared terminals?

10. What does the term *holding current* mean in reference to the operation of the SCR?

11. How does the removal of gate current affect the SCR's conduction once the device has been turned on?

12. How can the SCR be turned off once it has been triggered?

13. What is the relation between the magnitude of gate current and the magnitude of anode-cathode voltage during the time an SCR is being triggered?

14. How can an ohmmeter be used to test the general condition of an SCR?

15. What are some of the manufacturer's specifications that are significant to the operation of an SCR?

16. How does the diac differ from the Shockley diode?

17. How many semiconductor regions and junctions are associated with the diac's physical structure?

18. Why is the diac said to be like an AC switch?

19. What are some of the more significant manuacturer's specifications associated with the operation of the diac?

20. How do the diac and triac differ in physical and operating characteristics?

21. Why is the triac said to be like two SCRs connected in parallel?

22. What are some of the more important manufacturer's specifications concerning the operation of the triac?

23. How many semiconductor materials and junctions are associated with the physical structure of the UJT?

24. What is the intrinsic standoff ratio characteristic of a UJT?

25. What conditions must exist for emitter–base 1 current to flow in the UJT?

26. Why can the emitter–base 1 current of the UJT be used to trigger another component or circuit?

27. Why is the UJT said to exhibit a negative resistance characteristic?

28. What are some of the more important manufacturer's specifications concerning the operation of the UJT?

29. How can a transistor be used as a triggering device?

30. What is a thyratron tube and how does it differ from a hot-cathode gaseous diode tube?

31. What is the purpose of the grid of a thyratron tube?

32. What is the relation between grid voltage and plate voltage when turning the thyratron on?

33. What is an ignitron and how does it differ from a thyratron?

34. Why must all gaseous tubes be operated with an external load?

35. What polarity of voltage is applied to the ignitron's ignitor to cause the tube to ionize?

36. Why are all gaseous tubes said to be rectifiers?

PROBLEMS

1. A UJT exhibits an internal emitter–base 2 resistance of 4000 Ω and emitter–base 1 resistance of 500 Ω. What is its intrinsic standoff ratio?

2. If the UJT in problem 1 had a base-to-base voltage of 20 V, approximately what emitter voltage would cause emitter–base 1 current?

SUGGESTED LABORATORY ACTIVITIES

1. Using a Shockley diode, resistor, and the necessary electrical equipment, construct the circuit shown in Figure 5–31 and determine the forward switching voltage and holding current of the diode.

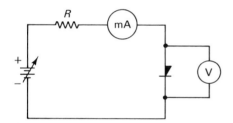

FIGURE 5–31   Shockley diode circuit.

2. Using an SCR, resistors, and the necessary electrical equipment, construct the circuit shown in Figure 5–32 and determine the gate cur-

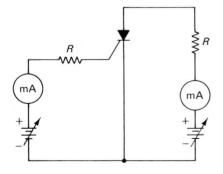

FIGURE 5–32   SCR circuit.

rent required to turn on the SCR when its anode-cathode voltage is 20, 30, and 40 V. Likewise determine holding currents for these anode-cathode voltages.

3. Using an ohmmeter, cause an SCR to trigger and hold conduction by jumping and removing the gate connection.

4. Using a diac, resistor, and the necessary electrical equipment, construct the circuits illustrated in Figure 5–33 and record the diac's breakover voltages.

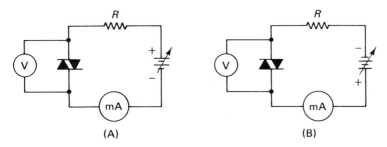

(A)   (B)

FIGURE 5–33   Diac circuits.

5. Using a triac, resistors, and the necessary electrical equipment, construct the circuits in Figure 5–34 to indicate how the triac can be triggered in both directions.

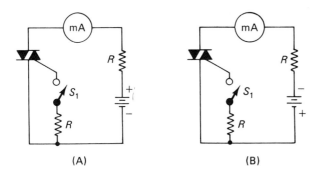

(A)   (B)

FIGURE 5–34   Triac circuits.

6. Using a UJT, resistors, and the necessary electrical equipment, construct the circuit illustrated in Figure 5–35 and determine the emitter voltage required to cause emitter–base 1 current.

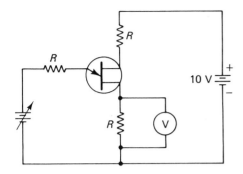

FIGURE 5–35   UJT circuit.

7.   Using a transistor, resistors, a capacitor, and the necessary electrical equipment, construct the circuit shown in Figure 5–36 to illustrate the transistor's triggering ability.

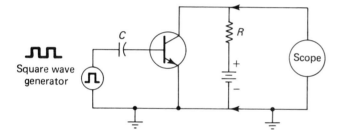

FIGURE 5–36   Transistor circuit.

8.   Using the thyratron and components supplied by your instructor, construct the circuit illustrated in Figure 5–37 and determine the plate-cathode voltage required to fire the thyratron when its grid voltage is − 1, − 2, and − 3 V.

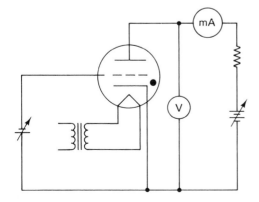

FIGURE 5–37   Thyratron circuit.

CHAPTER 6
Triggered Control Device Circuit Applications

In Chapter 5 the characteristics of many triggered control devices were examined. These devices, along with appropriate circuitry, are used to control the action of current or to control the action of other similar devices that are part of a control system.

In this chapter many control circuits that depend upon one or more of these devices to achieve their purpose will be examined. Keep in mind that most triggered circuits are designed to turn a resistive or inductive load "on" or "off" and to control the amount of current that flows through a load. Triggered circuits that control such things as heating, lighting, and motor speed play a vital part in our industrialized society. Since many of these circuit applications involve time constant and waveshaping circuits, a discussion of these circuits is included.

TIME CONSTANTS

Time constant circuits have several applications in electronics. They are used in many industrial processes and computer and communications circuit design. Timing functions may be as simple as on-off control or as complex as elaborate sequential operations. The controlled element of a timing circuit could be a motor, a relay, a solenoid, a lamp, or some other circuit or device.

Time constant circuits use the properties of inductance or capacitance to operate as timing circuits to control load devices. Remember that inductance (L) opposes changes in current and capacitance (C) opposes changes in voltage. The reaction time of inductors and capacitors to oppose changes of current and voltage depends on resistance. The time (in seconds) for a capacitor or an inductor to react is called its *time constant*.

RL TIME CONSTANTS

In Figure 6–1A a resistor and an inductor are connected in series to a voltage source. Current rises from zero to maximum after a certain time period due to the counterelectromotive force (CEMF) of the magnetic field. The time required for the current to reach maximum is controlled by the values of R and L. Resistance opposes current flow and inductance opposes changes in current. The change in current occurs from its zero to maximum values. The time required for the current to reach about 63 percent of maximum is found by using the formula $t = L/R$. Time constant *(t)* is in seconds, L is in henries, and R is in ohms. After one time constant, current through the inductor is about 63 percent of maximum (see Figure 6–1B). After five time constants, current through the inductor is maximum.

Assume that the maximum current (I_{max}) in the circuit is 100 mA and the time required for the current to reach 63 percent of 100 mA is 1.0 s. The time required for the current to reach the maximum (100 mA) is 5 s $(t \times 5 = 1.0 \times 5 = 5 \text{ s})$.

The current flow through the inductor (I_L) varies after the circuit switch is turned on until it reaches its maximum value after five time constants. The values of current are as follows:

After 1 time constant:	**63% × 100 mA = 63 mA**
After 2 time constants:	**86% × 100 mA = 86 mA**
After 3 time constants:	**95% × 100 mA = 95 mA**
After 4 time constants:	**98% × 100 mA = 98 mA**
After 5 time constants:	**100% × 100 mA = 100 mA**

Thus, after five time constants (5 s for this example), I_L is equal to its maximum value. The buildup of I_L in a RL circuit is called its *charging* condition. Charging occurs until I_L stabilizes at its maximum value, as shown in Figure 6–1B.

RL circuits also have a *discharge* condition. When the switch of Figure 6–1A is opened, I_L will reduce at a rate determined by the circuit time constant. Assume that a wire is placed across the points where the power source was connected at the same time the switch is turned off. The current through the inductor (I_L) will discharge, as shown in Figure 6–1C, until it stabilizes. The values of current during discharge are as follows:

After 1 time constant:	**37% of 100 mA = 37 mA**
After 2 time constants:	**14% of 100 mA = 14 mA**

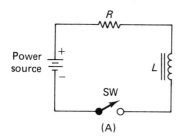

(A)

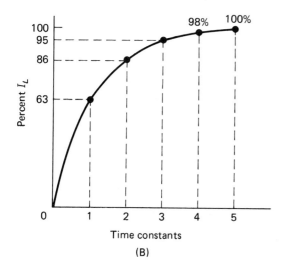

(B)

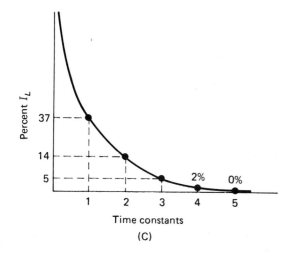

(C)

FIGURE 6–1   *RL* time constant circuit. (A) Circuit diagram.
(B) Time versus current curve—charging. (C) Time versus current curve—discharging.

After 3 time constants:	**5% of 100 mA = 5 mA**
After 4 time constants:	**2% of 100 mA = 2 mA**
After 5 time constants:	**0% of 100 mA = 0 mA**

RC TIME CONSTANTS

A resistor and capacitor are connected in series to a voltage source in Figure 6–2A. The time for the voltage across the capacitor to reach maximum is controlled by the values of R and C. Resistance opposes current flow in the circuit, while capacitance opposes changes in voltage. The change in voltage across the capacitor is from zero to maximum. The amount of time for a capacitor to charge to about 63 percent of the applied voltage is found by using the formula $t = R \times C$. Time (t) is in seconds, R is in ohms, and C is in farads.

After one time constant, the voltage across the capacitor is about 63 percent of the source voltage. Figure 6–2B shows a time versus voltage charging curve. In five time constants, the voltage across the capacitor approximately equals the source voltage. The time for the capacitor of Figure 6–2A to charge to 63 percent of the source voltage is one time constant, or 3 s. The time for the voltage across the capacitor to equal the source voltage is 15 s ($t \times 5 = 3 \times 5 = 15$ s).

The voltage across the capacitor (E_C) in an RC circuit increases until it reaches the source voltage value. When the switch in Figure 6–2A is closed, E_C will equal approximately 12 V after five time constants. The values of voltage across the capacitor are as follows:

After 1 time constant:	**63% of 12 V = 7.56 V**
After 2 time constants:	**86% of 12 V = 10.32 V**
After 3 time constants:	**95% of 12 V = 11.40 V**
After 4 time constants:	**98% of 12 V = 11.76 V**
After 5 time constants:	**100% of 12 V = 12.00 V**

Thus, after five time constants (15 s for this example), E_C is approximately equal to the source voltage. The buildup of voltage across the capacitor is called its *charging* condition, as shown in Figure 6–2B.

The *discharging* of the capacitor is similar to its charging action (see Figure 6–2C). When the source voltage is removed, the capacitor discharges about 63 percent of the source voltage in one time constant. In five time constants a capacitor discharges 100 percent of its voltage. When the switch of Figure 6–2A is opened, E_C will reduce at a rate determined by the circuit time constant. Assume that a wire is placed across the points where the power source was connected at the same time the switch is turned off. The voltage across the capacitor (E_C) will

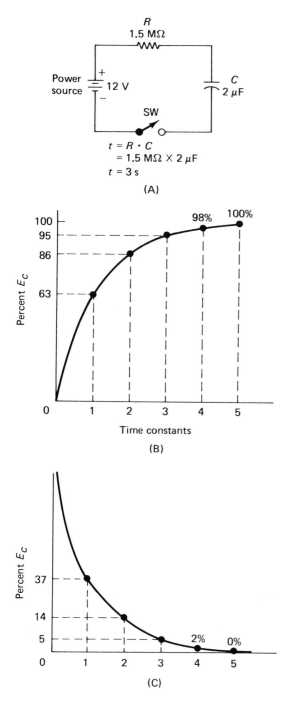

FIGURE 6–2 *RC* time constant curve. (A) Circuit diagram. (B) Time versus voltage curve—charging. (C) Time versus voltage curve—discharging.

discharge, as shown in Figure 6–2C, until it reaches minimum value. The values of capacitor voltage during discharge are as follows:

After 1 time constant: 37% of 12 V = 4.44 V
After 2 time constants: 14% of 12 V = 1.68 V
After 3 time constants: 5% of 12 V = 0.60 V
After 4 time constants: 2% of 12 V = 0.24 V
After 5 time constants: 0% of 12 V = 0.00 V

UNIVERSAL TIME CONSTANT CURVES

The time for current changes in inductive circuits and voltage changes in capacitive circuits is similar. The curves shown in Figure 6–3 are called *universal time constant curves*. The vertical axis is marked as percent of capacitor voltage (E_C) or percent of inductor current (I_L). The horizontal axis is marked in time constant values. Sometimes the charging curve is called the *rise time* and the discharging curve is called the *decay time*.

These curves are useful since they contain the voltage or current values for *RC* and *RL* circuits to be determined at any given time. For instance, the charge on a capacitor of a series *RC* circuit after 2.5 times constants can be estimated easily (see Figure 6–4). Since $t = R \times C = 15$ s for the circuit, the charge on the capacitor (E_C) after 2.5 time constants is between 8.6 and 9.5 V. A close estimation of voltage across the capacitor may be found by interpolation of values as follows:

$$E_C = 8.6 + (9.5 - 8.6)/2$$
$$= 8.6 + 0.45$$
$$= 9.05 \text{ V}$$

The charge on a capacitor after a given period of time can also be estimated on a universal time constant curve. For instance, after 25 s what is the approximate charge on the capacitor of Figure 6–4? This is a difference of 2.3 V. The value of 25 s is 10 s after the first time constant (25 s = 15 + 10). The time is then $^{10}\!/_{15}$, or $^{2}\!/_{3}$, of the distance from the first time constant (15 s). The approximation of E_C after 25 s is as follows:

$$E_C = 6.3 + (^{2}\!/_{3} \times 2.3)$$
$$= 6.3 + 1.53$$
$$= 7.83 \text{ V}$$

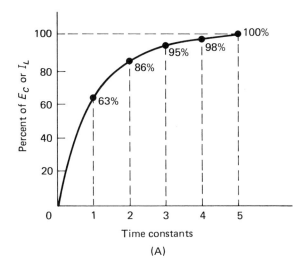

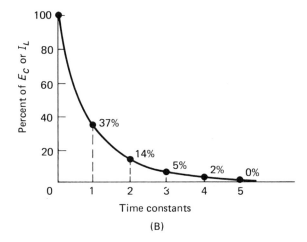

FIGURE 6–3   Universal time constant curves. (A) Charging or
rise time curve. (B) Discharging or decay time curve.

It should be pointed out that values along the logarithmic universal
time constant curve using the methods outlined above are approximations. The exact values may be obtained using calculus procedures and
natural logarithms. Approximations such as those explained are commonly used for electronic circuit design where extremely precise values
are not necessary.

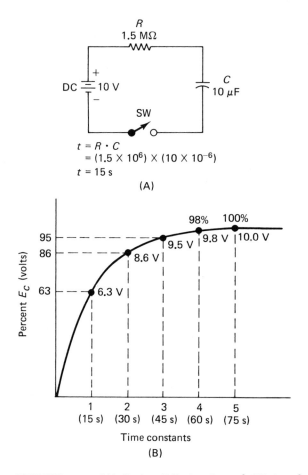

FIGURE 6–4   (A) Series *RC* circuit and (B) its charging time constant curve.

WAVESHAPING CONTROL WITH TIME CONSTANT CIRCUIT

A circuit that is designed to change the shape of a waveform is very essential in electronics today. The type of waveshape achieved is primarily based on the shape of the applied signal and the characteristics of the shaping circuit. *Sine, square,* and *sawtooth* waves are the basic input waves to be processed. Square and sawtooth waveforms are shown in Figure 6–5. Waves may be of the periodic type or have a pulsing characteristic. A *periodic wave* continually changes value within a given

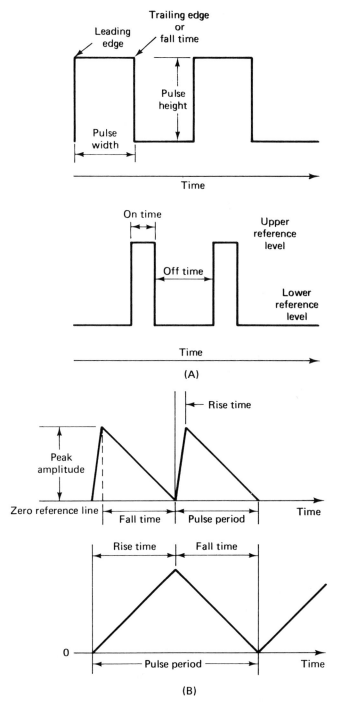

FIGURE 6–5 (A) Square or rectangular waveforms and (B) sawtooth waveforms.

time frame. *Pulse waveforms* occur momentarily at intervals. *Repetitive* pulses occur at definite time periods. *Transient pulses* occur randomly and are the result of a switching action or some other electrical change. Waveshaping circuits may be used to minimize transient pulse conditions so that they will not cause irregular control of a circuit.

Numerous devices and components have been developed and are being used to achieve various waveshaping operations. *Waveshaping* is considered to be an altering process that changes such things as the basic form of the wave or its conduction time or causes a variation in its amplitude.

A rather simple method of altering the basic form of a wave is to apply it to an *RC* time constant circuit. Two distinct form changes can be produced according to the *RC* component combination selected. With a square or rectangular wave applied, a differentiated or integrated waveshape may be produced. Refer to the waveshaping circuits of Figure 6–6.

DIFFERENTIATOR CIRCUITS

The *differentiator* circuit of Figure 6–6A has a square-wave input signal applied to a capacitor, and its output is developed across a resistor. Any abrupt change in the input signal will cause a pronounced change or dif-

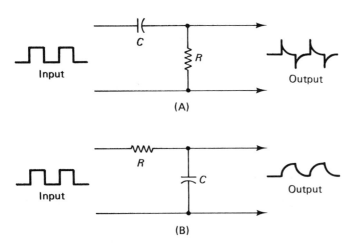

FIGURE 6–6 *RC* waveshaping circuits. (A) differentiator circuit. (B) Integrator circuit.

ference in the output voltage developed across R. The term *differentiator* is used to describe the changing output of this circuit.

When the leading edge of an applied square wave rises quickly, it causes the capacitor of a differentiator to charge immediately. The series-connected resistor of this circuit "sees" a current that is identical to that produced by the capacitor. This current rises sharply to a peak and then drops off very quickly when the capacitor becomes charged. Current passing through R develops an output voltage that conforms to the charging current of the capacitor.

When the trailing edge of an applied square wave is reached, it causes an abrupt drop in the applied voltage to the capacitor. The electrostatic charge on the capacitor now begins to discharge back through the resistor. This current is in a direction opposite to that of the original charging current. The resulting current through R causes a corresponding output voltage to be developed. The polarity of this voltage is opposite to that of the charging cycle. The output voltage will then drop below the zero reference line.

When a differentiator has a short time constant with respect to the applied square wave, the output will have only sharp positive and negative spikes. A longer time constant, however, will cause the trailing edge of the output to be extended somewhat. This indicates that the discharge action of the capacitor has been prolonged. The *RC* time constant of a differentiator with respect to the *pulse repetition rate (PRR)* of the square-wave input largely determines the form of the output of this circuit. The reciprocal of the time that it takes for a wave of this type to repeat itself is called the pulse repetition rate, or PRR. This term compares to the frequency of a sine wave, which is expressed in hertz (Hz).

INTEGRATOR CIRCUITS

The *integrator* of Figure 6–6B is a waveshaping circuit that has the input signal applied to a resistor and the output developed across a capacitor. The value of the resistor in this case tends to restrict the charging time of the capacitor. The voltage developed across C therefore has a gradual rise time and fall time when a square wave is applied.

The output of an integrator can be altered by changing the PRR of the input or by changing the time constant of the *RC* network. Time constants longer than the on time of a square wave tend to cause it to have a sawtooth appearance. A shorter time constant tends to cause a slope on the leading and trailing edge followed by straight-line areas. Integrators are primarily used to sum, or total, signal values that are variable.

SAWTOOTH GENERATOR CIRCUITS

Sawtooth generators are used where a signal must have a slow rise time and a rather rapid fall time. A sawtooth wave can be produced very easily by connecting a resistor and capacitor in series with a DC energy source. The charging and discharging action of the capacitor then can be controlled. A circuit of this type is often called a *relaxation oscillator*. When the switching device is conducting, the circuit is said to be in an *active state*. The nonconducting condition of this device is then referred to as the *relaxation state*. In practice this oscillator switches back and forth between the active and relaxation states.

The circuit shown in Figure 6–7 is a type of relaxation oscillator or sawtooth generator circuit. The *RC* time constant of the circuit determines the flash rate of the lamp and the pulse repetitive rate of the circuit. When DC voltage is initially applied to the circuit, the capacitor will charge to a value equal to the ionization voltage of the neon lamp. When the neon lamp ionizes or becomes conductive, its resistance changes from near infinity to a very low value. In this respect, the neon lamp acts as a switch which opens and closes at a speed based on the circuit time constant. Once the neon lamp becomes conductive, the capacitor will discharge very rapidly through the lamp. The decreased charge on the capacitor will then cause more charge voltage to be drawn from the voltage equal to the ionization voltage of the neon lamp. This process will repeat itself at the time rate determined by the circuit's *RC* time constant. Increasing the value of *R* or *C* will cause a longer rise time of the output signal.

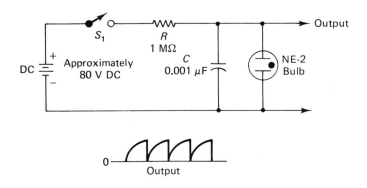

FIGURE 6–7   Sawtooth generator circuit.

SHOCKLEY DIODE CIRCUITS

Recall that the Shockley diode is a PNPN diode that is forward biased and conducts when its forward switching voltage is large enough to cause it to be triggered, or turned on. Because of these unique characteristics, the Shockley diode is used in oscillator and time delay circuits. The Shockley diode is also used as a trigger device that turns on the SCR. Figure 6–8 illustrates the Shockley diode being used in an oscillator circuit.

An oscillator is an electronic circuit designed to create electrical waveforms or pulses at a rate of a few cycles or pulses per second, or many millions of cycles or pulses per second. (Oscillators will be more fully discussed in Chapter 9.) In Figure 6–8 voltage across C_1 equals the forward switching voltage of D_1 when D_1 is turned on. This allows C_1 to discharge through D_1 and R_2. The discharge of C_1 through R_2 causes the voltage across R_2 to rise very rapidly from zero to a peak and drop back to zero as the voltage across C_1 decreases to zero. The decrease in voltage across C_1 allows the Shockley diode to be turned off, assuming that the combined resistance of C_1 and R_2 is large enough to limit the current below the diode's holding current level. This allows C_1 to again charge through R_1 until the voltage across C_1 is large enough to turn the diode on again. This causes the entire cycle to be repeated. These cycles continue over and over as long as all remains unchanged. The number of pulses per second produced across R_2 is determined by the RC time constant of R_1 and C_1.

If only one pulse output is desired, the values of R_1 and R_2 are carefully chosen to allow the Shockley diode to remain on once it has been triggered by the discharging capacitor. That is, the current flow allowed by R_1 and R_2 after the diode is turned on is larger than the diode's minimum holding current. The output across R_2 would appear as illustrated in Figure 6–9.

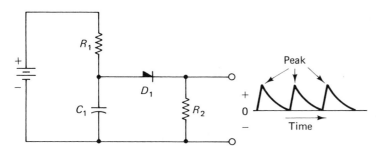

FIGURE 6–8   Shockley diode oscillator.

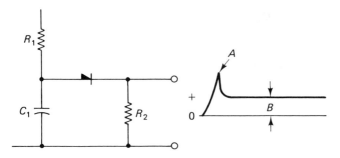

FIGURE 6–9   Shockley diode—one pulse output.

Point A represents the sharp rise in the voltage drop across R_2 when the diode is turned on and C_1 discharges. Since the diode remains on, the voltage across R_2 does not drop to zero but to some amount greater than zero, as illustrated at point B. This arrangement is used in time delay applications. The time difference between when voltage is applied to the circuit and when the pulse is produced across R_2 is controlled by the values of R_1 and C_1.

Figure 6–10 illustrates the Shockley diode being used to trigger an SCR. In this circuit the SCR is used as a switch to turn on the load after a certain time delay. When S_1 is closed, C_1 begins to charge toward the source voltage through R_1. After a certain amount of time has elapsed, the voltage across C_1 is great enough to turn on D_1. When D_1 is turned on, C_1 discharges through D_1 and R_2 causing a sharp spike of voltage drop across R_2. This spike is positive going and makes the gate of the SCR more positive than the SCR's cathode, resulting in SCR gate current.

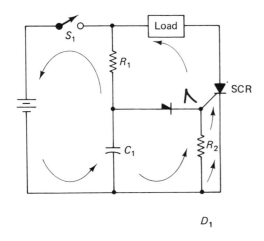

FIGURE 6–10   Shockley diode triggering an SCR.

The gate current triggers the SCR, allowing current to flow through the load. The time difference between the closing of S_1 and activation of the load is dependent upon the RC time constant of R_1 and C_1.

SCR CIRCUITS

The SCR may be used as a DC switch employing either a separate gate voltage source or a common anode-cathode-gate voltage source to achieve conduction. Figure 6–11 illustrates both types of circuits. The circuit in Figure 6–11A employs a separate gate voltage source. When S_1 is closed, the anode of the SCR is made positive and the cathode is made negative by virtue of V_1. The SCR will not conduct until S_2 is closed, placing a positive potential from V_2 on the SCR's gate. With S_2 closed, gate current flows, triggering the SCR. This allows current to pass through the SCR and load.

The circuit in Figure 6–11B employs one voltage source (V_1) to provide both the anode-cathode voltage and the gate voltage. When S_1 is closed, a voltage of appropriate polarity is applied across the SCR. The SCR will not conduct until S_2 is closed, allowing current to flow through R_1. R_1 acts as a voltage divider and, if adjusted properly, provides a positive voltage to the gate of the SCR. This causes gate current, turning the SCR on and allowing current to flow through the load.

The SCR has many more applications as a rectifier rather than as a DC switch. Figure 6–12 illustrates the SCR with a 60-Hz AC voltage applied to its anode-cathode. This circuit uses a separate DC gate voltage

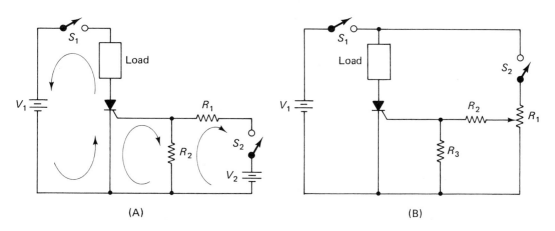

(A)   (B)

FIGURE 6–11   SCR as a DC switch.

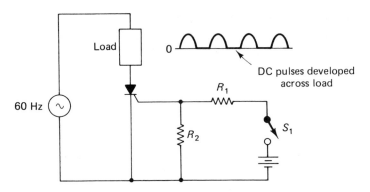

FIGURE 6–12   SCR as a DC triggered rectifier.

source. Since the SCR has AC voltage applied across its anode-cathode, its anode is alternately positive and negative with respect to its cathode at a rate of 60 Hz. Recall that an SCR only conducts when its anode is positive with respect to its cathode. This means that with S_1 closed the SCR's gate is made positive, but the SCR will only conduct during the half of the AC cycle that causes its anode to be positive. Thus, the SCR acts as half-wave rectifier and turns on and off 60 times per second. The current flow through the load is a series of DC pulses at a rate of 60 per second, assuming the gate potential is great enough to cause a sufficient amount of gate current to trigger the SCR as soon as its anode becomes positive. The DC pulses developed through the load are shown in Figure 6–12.

Assume that the gate potential is not great enough to trigger the SCR as soon as its anode becomes positive but is great enough to trigger the SCR as its positive anode voltage rises toward its peak. This delays the triggering of the SCR and causes it to conduct less than a full half-cycle.

The terms *cycle* and *circle* mean the same. Thus, the number of degrees of conduction allowed the SCR for any given gate current may be discussed. Both the cycle and circle exhibit 360° from start to finish, as illustrated on Figure 6–13. Thus, the most an SCR will conduct is 180° of the 360° since its anode is positive only during this half of the cycle.

Figure 6–14 illustrates how the level of gate current affects the triggering of the SCR when AC voltage is applied across its anode-cathode. Recall from Chapter 5 that the SCR is triggered by a certain amount of gate current for a given amount of anode-cathode voltage. The larger the gate current, the smaller the anode-cathode voltage must be in order to trigger the SCR. Likewise, the smaller the gate current, the larger the anode-cathode voltage must be to turn the SCR on. Figure 6–14 illustrates this effect when an AC voltage is applied across the SCR.

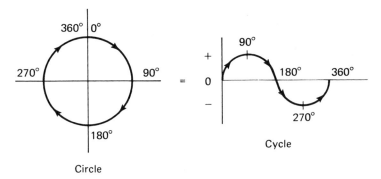

FIGURE 6–13   Circle-cycle analysis.

As illustrated in Figure 6–14A, the voltage across the SCR is an instantaneous voltage, rising from 0 V, peaking at 80 V at about 90°, and falling back to 0 V at 180°. At 180° the voltage reverses polarity and renders the SCR nonconductive for any amount of gate current. As illustrated to the left of the SCR voltage, a gate current ranging from 10 to 1 mA may be applied to the SCR when the SCR anode voltage is positive. Thus, when the gate current is 10 mA, the SCR is triggered as soon as its anode becomes positive (point M) and allows current to flow through the load for 180° (a full half-cycle) as illustrated in Figure 6–14B.

When the gate current is 7 mA, the anode voltage must rise to about 30 V before the SCR will trigger (point N). This causes about a 30° triggering delay. The SCR is not turned on until after about 30° of the anode voltage cycle is complete. This results in current flow through the load for only 150° of the 180° cycle (180° half-cycle minus first 30° when the SCR is off equals 150° of remaining conduction, or 180° − 30° = 150°), as illustrated in Figure 6–14C.

If the SCR's gate current is 4 mA, the instantaneous anode voltage must equal 60 V before it is triggered (point D). This results in a 60° triggering delay. As shown in Figure 6–14D, load current starts to flow at about the 60° point of the 180° cycle and continues for the remainder of the cycle. In this instance load current flows for 120° of the complete 180° cycle (180° − 60° = 120°).

Finally, as the SCR's gate current is 2 mA, the instantaneous anode voltage must be 80 V, which is the peak voltage in this case, before the SCR triggers (point P). This causes a 90° triggering delay in the 180° half-cycle and results in load current flow for 90° of the 180° half-cycle, as illustrated in Figure 6–14 (180° half-cycle minus 90° delay equals 90° conduction, or 180° − 90° = 90°).

It is important to note that a delay in the triggering of the SCR controls the amount of load current. The longer the delay, the less the cur-

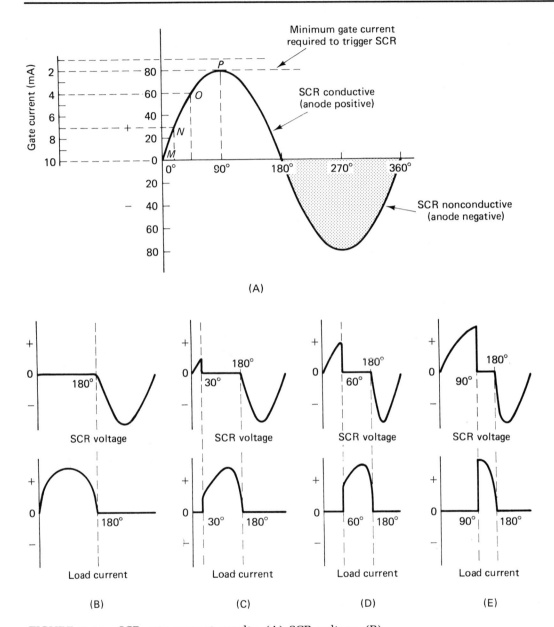

FIGURE 6–14   SCR gate current results. (A) SCR voltage. (B) Load current. (C) Load current. (D) Load current. (E) Load current.

rent flow through the load. The shorter the triggering delay, the more the current flow through the load.

Thus, the illustration shows that when the instantaneous anode voltage equals 80 V peak, a gate current between 10 and 2 mA causes the SCR to conduct for all or a portion of the 180° half-cycle. If the gate current is equal to or more than 10 mA, the SCR triggers as soon as its anode is positive. If the gate current is less than 2 mA, the SCR will not trigger as long as its instantaneous peak voltage is no greater than 80 V. Since the anode voltage decreases to zero after reaching its instantaneous peak, DC gate voltage and current can only delay the triggering of the SCR by a maximum of about 90°.

A more common control circuit appears in Figure 6–15. Here, the same AC source supplies the SCR's anode-cathode voltage, as well as the gate-cathode voltage that triggers the SCR. In this circuit D_1 acts as a half-wave rectifier and creates a pulsating direct current that is positive with respect to the SCR's gate. R_3 may be adjusted to cause the gate voltage, and thus current, to vary in amplitude. Since the anode-cathode and gate-cathode voltages are in phase, the proper adjustment of R_3 can bring about a 90° trigger delay in the conduction of the SCR.

Figure 6–16 illustrates the effect on the conduction of the SCR by causing the amplitude of the gate current to vary. From Figure 6–16A it can be seen that when the gate current equals an instantaneous peak of 10 mA, the SCR will trigger almost immediately when its anode becomes positive (point M). This results in a voltage drop across the SCR as illustrated in Figure 6–16B and a resulting load current flow of 180° of the SCR voltage cycle.

When the gate current equals a peak of 6 mA, the triggering of the SCR is delayed by about 30° (point N). This causes an SCR voltage and load current as illustrated in Figure 6–16C. Likewise, when the gate cur-

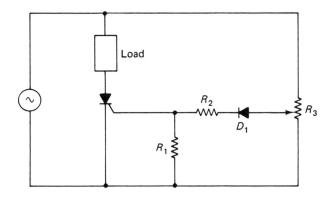

FIGURE 6–15   SCR as a controlled AC rectifier.

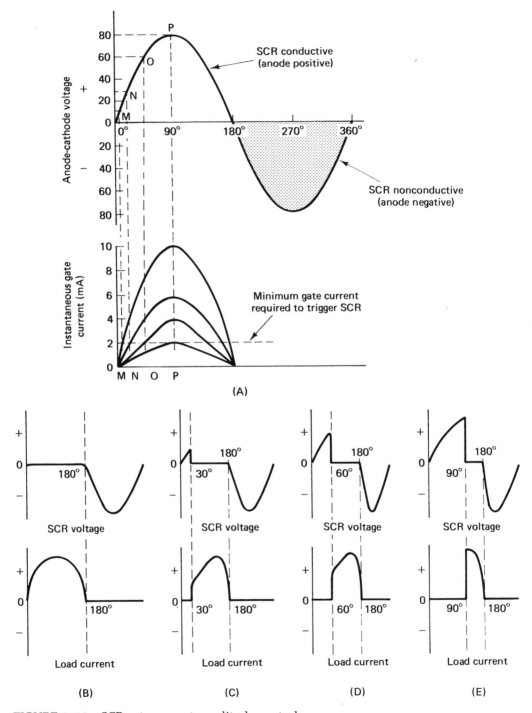

FIGURE 6–16   SCR gate current amplitude control.

rent is 4 mA, the anode-cathode voltage must be 60 V in order to cause the SCR to trigger (point D). This results in a 60° delay in its conduction causing the voltage and load current illustrated in Figure 6–16D.

Finally, when the gate current is decreased to 2 mA, the anode-cathode voltage must be 80 V if the SCR is to be triggered (point P). Since 80 V represents the peak voltage across the SCR, a peak current of 2 mA results in a 90° triggering delay, as illustrated in Figure 6–16E. No more than a 90° triggering delay can be achieved when the SCR's anode-cathode voltage and its gate voltage, and thus its current, are in phase.

If the phase of the pulsating DC gate voltage and current is electrically shifted to cause the gate potential to lag behind the SCR's anode-cathode voltage, a triggering delay from approximately 0° to 180° is realized. The circuit in Figure 6–17 allows the SCR to be triggered in this fashion. In this circuit the anode-cathode voltage is 180° out of phase with the gate-cathode voltage because of the voltage division caused by R_4 and R_5. The ratio of R_1 and X_L caused by L_1 can change the phase of the rectified gate-cathode voltage, causing it to become very nearly in phase with the anode-cathode voltage or to be very nearly 180° out of phase. Since R_1 may be adjusted (changing the ratio between R_1 and X_L), the phase relation is likewise variable resulting in the voltages illustrated in Figure 6–18.

It can be seen that when the anode-cathode and gate-cathode voltages are in phase, the anode and gate are made positive at the same instant, causing the SCR to be immediately triggered and to conduct for a full 180° of its cycle. If the anode-cathode and gate-cathode voltages are 180° out of phase, the SCR would never be triggered because when its anode voltage is positive, its gate is negative and vice versa. By shifting

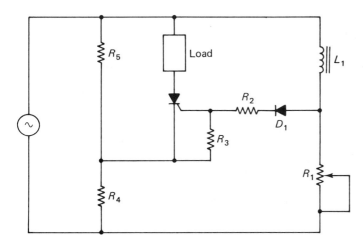

FIGURE 6–17   SCR control—0°–180°.

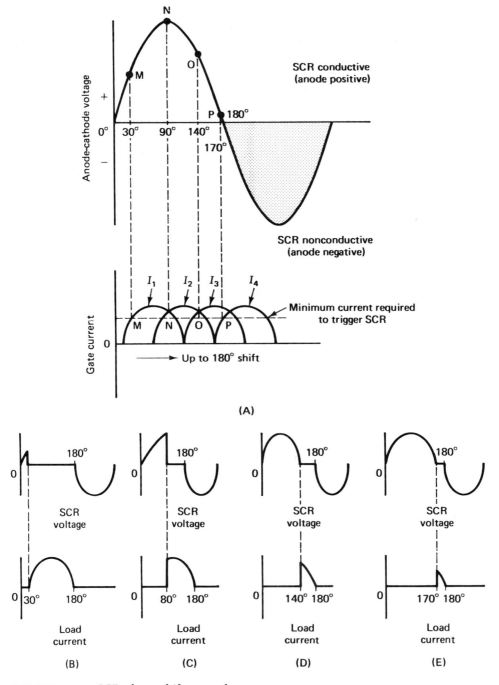

FIGURE 6–18   SCR phase shift control.

the gate potential between being in phase and being 180° out of phase with the anode potential, the SCR's conduction is controlled during the entire 180° of its cycle.

Assume that R_1 in Figure 6–17 is adjusted to cause the gate voltage and its current (I_1) to lag behind the anode voltage by 30°. As illustrated in Figure 6–18A, the triggering of the SCR is delayed to point M and results in the voltage and load current illustrated in Figure 6–18B. Likewise, if R_1 is adjusted to cause the gate potential to lag more, as illustrated by gate currents I_2, I_3, and I_4, the SCR is triggered until its anode voltage reaches points N, O, and P of Figure 6–18A. This results in the SCR voltages and load currents illustrated in Figure 6–18C–E. As shown, the greater the difference in the phase relation between the gate potential and the anode potential, the longer the SCR is delayed in turning on and the less current flows through the load. If the triggering action of the SCR is delayed by as much as 170°, as illustrated in Figure 6–18E, load current only flows for about 10° of the 180° cycle. If the load in the illustration becomes a DC motor, its speed is easily controlled by adjusting R_1. Likewise, if the load becomes the heating element of an electric furnace, its temperature is very easily controlled by simply adjusting R_1 of the illustrated circuit.

DIAC-TRIAC CIRCUITS

Recall from Chapter 5 that the diac is a two-lead, three-region (NPN) solid-state device that is described as an AC switch. It can be turned on, or triggered, to conduct in either direction when anodes 1 and 2 are made alternately positive and negative.

Likewise, recall that the triac is like two SCRs connected back-to-back in a parallel arrangement with a common gate. The gate potential of a triac may be either positive or negative, resulting in the triggering of the device when its anodes are either positive or negative. While the SCR is designed to control DC load currents, the triac is designed to control AC load currents.

The diac is most commonly used to turn on the triac, which in turn controls AC current to motors, heating elements, lights, and other loads.

The circuit illustrated in Figure 6–19 is most commonly used to control an AC current to a load. When S_1 is closed, an AC voltage is applied across the triac and R_1–C_1 of the circuit. C_1 begins to charge through R_1. When C_1 charges to a voltage that is large enough to trigger the diac (20–30 V), the diac becomes conductive and allows C_1 to discharge through the gate of the triac. The discharge of C_1 triggers the triac into conduction, which results in a current that activates the load.

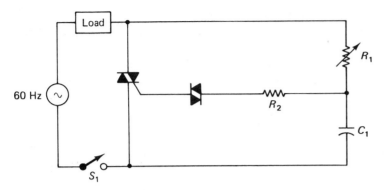

FIGURE 6–19   Diac-triac AC load control.

When the AC voltage reverses polarity, C_1 charges through R_1 in the manner previously described but to the opposite voltage polarity. As the voltage across C_1 increases, the diac is turned on and C_1 again discharges through the gate of the triac, triggering it into conduction. The resulting voltages across the triac and currents through the load are illustrated in Figure 6–20.

The delay in the triggering of the triac is controlled by the RC time constant of R_1 and C_1. By adjusting R_1, the time required for C_1 to charge to a voltage that will trigger the diac, and thus the triac, can be controlled. This in turn controls the amount of time that load current flows and thus the amount of load current. When R_1 is adjusted to some small

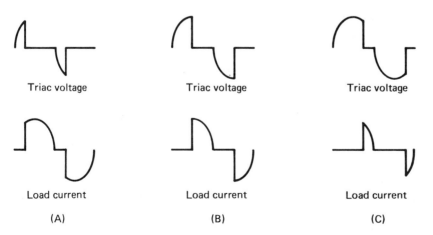

FIGURE 6–20   Controlled AC load current.

amount of resistance, C_1 will charge very quickly, triggering the diac and triac. This allows the triac to conduct almost 360° of the AC cycle, resulting in a large load current (Figure 6–20A). As the value of R_1 increases, C_1 charges more slowly, which results in an increased delay in the conduction of the triac and a lesser amount of load current (Figure 6–20B). An additional increase in the value of R_1 causes the circuit waveforms illustrated in Figure 6–20C.

UJT CIRCUITS

Recall from Chapter 5 that the unijunction transistor (UJT) is a three-terminal, solid-state device that has one junction. This junction is formed between the emitter (P material) and the conductive path (N material) provided between bases 1 and 2. The more interesting characteristics of the UJT are its *intrinsic standoff ratio* and its *negative resistance*.

The UJT is used in time delay, oscillator, and trigger circuits. Figure 6–21 illustrates the UJT used in an oscillator circuit. When S_1 is closed, C_1 charges through R_3. As the voltage across C_1 reaches or slightly exceeds the UJT's intrinsic standoff ratio voltage, the emitter becomes forward biased and conductive. This allows C_1 to discharge through R_2 and the base 1–emitter junction. The discharge of C_1 causes a sharp increase in the voltage drop across R_2 illustrated as output in Figure 6–21. When C_1 becomes discharged, the emitter is no longer forward biased. It then becomes nonconductive, and the voltage drop across R_2 returns to its normal level. C_1 charges again through R_3 and the

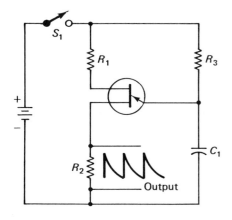

FIGURE 6–21 UJT oscillator.

entire cycle is repeated. The frequency of the resulting output across R_2 in Figure 6–21 is determined by the source voltage, the RC time constant of R_3 and C_1, the value of R_2, and the intrinsic standoff ratio of the UJT.

A variation of the circuit illustrated in Figure 6–21 becomes a tone generator, as shown in Figure 6–22. In this circuit the values of R_1, R_2, and R_3 are different. When S_1 is closed, C_1 charges, triggering the emitter of the UJT. C_1 discharges through the speaker coil and the emitter–base 1 junction. After C_1 discharges, the emitter is nonconductive and allows C_1 to again charge through R_1. The entire cycle is repeated, resulting in an audible tone from the speaker. The frequency of the tone is controlled by the RC time constant of R_1 and C_1.

If S_1 is open and S_2 is closed, a new time constant is in effect. This causes the speaker to produce a second tone of a different frequency. Closing any, all, or any combination of S_1, S_2, and S_3 causes the speaker to produce as many as seven different tones.

When used in time delay circuits, the UJT works in essentially the same way as in the oscillators previously described. The most important consideration is the selection of the values of R and C which control the time when the UJT is triggered.

Figure 6–23 illustrates a UJT used in a time delay circuit to trigger an SCR. With S_1 closed, C_2 begins to charge through R_1. When the voltage across C_1 is great enough to forward bias the emitter of the UJT, the emitter–base 1 junction becomes conductive, providing a discharge path for C_1. The discharge of C_1 causes a sharp increase in the voltage drop across R_3. This sharp voltage pulse is positive going and is applied to the gate of the SCR. The SCR is triggered "on" and the current flows through the load. The time delay between the closure of S_1 and the triggering of the SCR is determined by the RC time constant of R_1 and C_1.

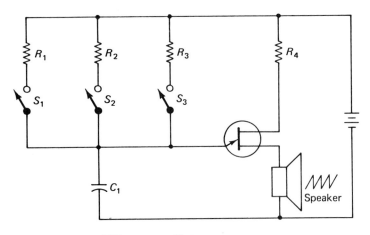

FIGURE 6–22   UJT tone oscillator.

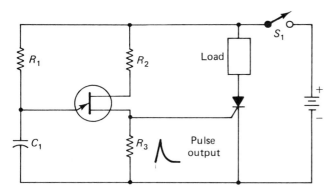

FIGURE 6–23   UJT time delay SCR trigger.

TRANSISTOR TRIGGER CIRCUITS

The transistor is a three-terminal (base, emitter, collector), three-region (NPN or PNP), solid-state device. Its conduction is controlled by properly biasing its two junctions, as discussed in Chapter 4.

The transistor is used as a triggering device where low voltage-current *sensing circuits* are to trigger high voltage-current loads. One example of this application appears in Figure 6–24. In this circuit the temperature sensor exhibits a positive temperature coefficient. That is, as its temperature increases, its resistance increases. Likewise, as its temperature decreases, its resistance decreases.

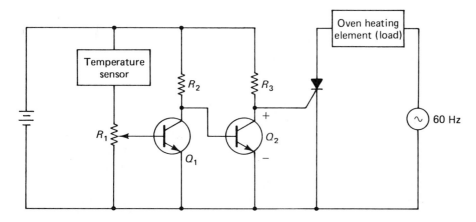

FIGURE 6–24   Transistors triggering SCR.

Assume that the sensor is placed inside an electric oven several yards from the circuit. Further assume that the oven is heated to a desired temperature which must be maintained. R_1 is then adjusted to cause transistor Q_1 to be forward biased and conduct emitter-collector current. Its conduction causes a voltage drop across its emitter-collector which provides the positive forward-bias voltage to the base of transistor Q_2. The conduction of Q_2 and the resulting emitter-collector voltage provide the positive gate voltage-current to the SCR.

Since the resistance of R_1 ultimately determines the triggering point of the SCR, it must be very carefully adjusted to allow the emitter-collector voltage of Q_2 to be just below a voltage value that triggers the SCR. In doing so, this sensing circuit causes a minimum temperature to be maintained within the oven.

Assume the temperature in the oven decreases for some reason, resulting in a decrease in the resistance of the temperature sensor. This reduces the resistance of the branch of the circuit that contains R_1, allowing more current flow through R_1. This causes an increase in the voltage drop across R_1, making the base of Q_1 more positive, reducing its emitter-collector resistance and causing more emitter-collector current to flow through Q_1. This results in less emitter-collector voltage drop across Q_1. This reduction in emitter-collector voltage across Q_1 reduces the positive potential at the base of Q_2. This causes an increase in the emitter-collector resistance of Q_2 and a reduction in the emitter-collector current of Q_2. The emitter-collector voltage rises as a result, causing the gate of the SCR to become more positive. The SCR is triggered into conduction, causing a pulsating DC current to flow through the heating element of the oven. This pulsating DC current warms the heating element, causing the temperature in the oven to rise. The increase in oven temperature results in the temperature of the oven's temperature sensor being returned to its original level.

When the temperature of the sensor is returned to its original level, its resistance is returned to its original level. This causes the current in the branch of the circuit that contains R_1 to return to its original level, reducing the conduction of Q_1. The reduction in the conduction of Q_1 results in an increased emitter-collector voltage across Q_1. This causes the base of Q_2 to become more positive and decreases the emitter-collector resistance of that transistor. This causes the emitter-collector voltage across Q_2 to decrease, resulting in a decreased SCR gate voltage.

Since the SCR is conducting at a rate of 60 pulses per second by virtue of its AC source voltage, it is turned off when its gate potential drops below the minimum trigger level. The SCR remains off, causing current to no longer flow through the oven's heating element. A decrease in the temperature of the sensor again brings about the chain of events previously described. Thus, a few milliamperes control the action of several amperes of load current.

Figure 6–25 illustrates a transistor controlling the triggering of an

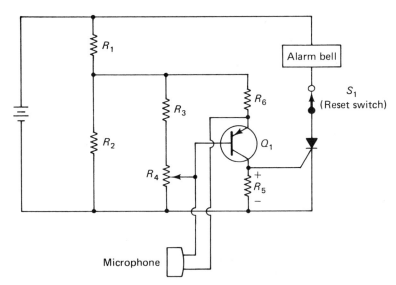

FIGURE 6–25   SCR sound sensor.

SCR in a sound-sensing security system circuit. In this circuit R_4 is adjusted to control the base current of Q_1 and causes the emitter-collector voltage, and thus the gate-cathode voltage, of the SCR to be just below the SCR's gate-triggering potential.

Since the microphone is a *transducer* (a device that changes one form of energy into another form of energy), any sound produced within its range is changed to a small voltage that is applied between the base and emitter of Q_1. This small voltage that results from a sound causes Q_1 to be slightly more forward biased during the instant that the sound is produced. This causes Q_1 to conduct more emitter-collector current, increasing the voltage drop across R_5, the gate-cathode voltage. This increase in gate potential triggers the SCR, causing the alarm bell to ring.

Since the anode-cathode voltage of the SCR is steady direct current, the SCR remains conductive once it is triggered. The alarm bell continues to ring until someone removes the SCR's anode potential by momentarily opening S_1. The system is then prepared to sense additional sounds made by intruders or malfunctioning equipment once S_1 is closed again.

THYRATRON CIRCUITS

The thyratron is a gaseous half-wave rectification tube that contains a filament or heater, cathode, plate, and control grid. Like the SCR, its triggering time can be delayed up to 180° of its conduction cycle with the proper control grid voltage. Unlike the SCR, the voltage that delays or controls the thyratron's conduction must be negative with respect to its cathode.

Figure 6–26 illustrates one method of supplying grid voltage that delays the conduction of the thyratron up to 90°. The circuit employs a variable DC voltage supply to provide a negative voltage to the control grid of the thyratron. Adjusting the DC voltage supply affects the conduction of the tube and causes the waveforms illustrated in Figure 6–27.

As illustrated in Figure 6–27A, − 4 V on the grid of the tube will not allow the thyratron to conduct until its plate voltage is about 50 V (point M). This delays the conduction of the tube for about 30° of its 180° half-cycle and results in the waveforms illustrated in Figure 6–27B.

If the grid voltage is − 8 V, the tube's conduction is delayed until its plate voltage is 150 V (point N). This causes a delay in tube conduction of about 60° and causes the waveforms illustrated in Figure 6–27C.

If as much as − 12 V exist between the grid and cathode of the tube, the thyratron's conduction is delayed by 90° (point O). This causes the waveforms appearing in Figure 6–27D. It can be seen that when the grid voltage is − 12 V, the plate voltage must rise to 250 V before the tube will turn on. Since 250 V is the peak voltage across the tube, it can be seen that a steady DC grid voltage only delays the tube's conduction by 90°. More than − 12 V on the grid causes the tube to remain off at all times. Less than − 12 V on the tube's grid allows it to conduct when the plate voltage is less than 250 V.

A more common method of controlling the conduction of the thyratron is illustrated in Figure 6–28. In this circuit the grid voltage is al-

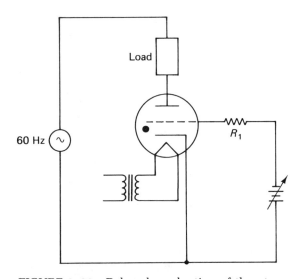

FIGURE 6–26   Delayed conduction of thyratron.

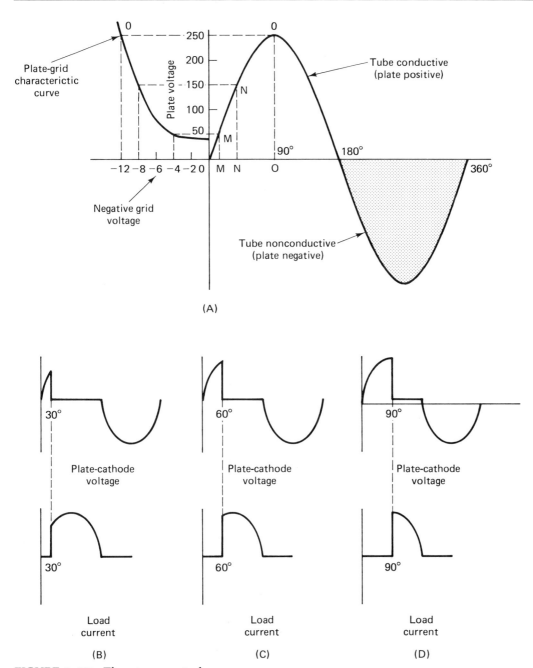

FIGURE 6–27 Thyratron control.

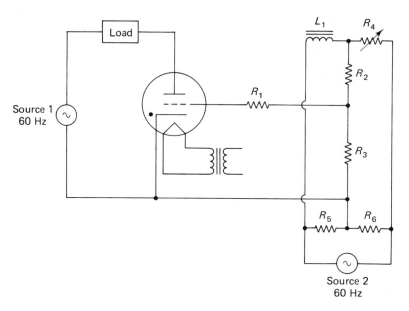

FIGURE 6–28   Phase shift thyratron control.

ternating current and is connected in such a way as to cause its negative half-cycle to lag the positive half-cycle of plate voltage. The amount of lag (0°–180°) is controlled by R_4 as its adjustment changes the ratio of inductive reactance (X_L) caused by L_1, and resistance provided by R_4. It should be pointed out that the two AC voltage sources provide voltages that are 180° out of phase, and that by adjusting R_4, the relation between these voltages is changed. Furthermore, the amplitude of the AC grid voltage must be greater than a voltage value that allows the thyratron to trigger when its plate voltage reaches its peak value. The amplitude of the grid voltage in the circuit illustrated in Figure 6–28 is controlled by the voltage divider network formed by R_2 and R_3.

Figure 6–29 illustrates the approximate effect on the thyratron's conduction as the phase of its grid voltage is shifted. In Figure 6–29A it can be seen that when the grid voltage and plate voltage are in phase, the tube will trigger as soon as its ionization potential is reached and there will be no delay in its conduction. A shift in the grid voltage, as shown in Figure 6–29B, delays the conduction of the tube by about 80° and causes the illustrated waveforms. A greater shift in grid voltage brings about a conduction delay of about 160° as illustrated in Figure 6–29C. The tube never conducts if the grid voltage is shifted to 180° out of phase with the plate voltage, assuming the amplitude of the grid voltage is properly chosen. This is illustrated in Figure 6–29D.

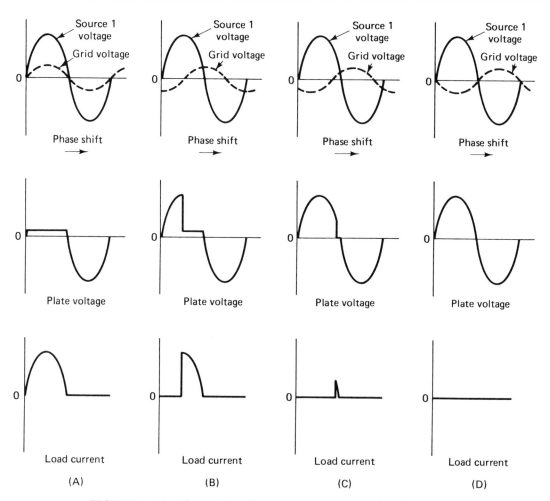

FIGURE 6–29   Phase shift effects on thyratron conduction.

As stated earlier in this text, the thyratron is rapidly being replaced by the SCR. Even though this is true, the thyratron still has limited application in the manufacturing and communications industries.

IGNITRON CIRCUITS

The ignitron is a gaseous rectifier that employs no heater but does employ an ignitor, cathode, and plate. Its cathode is usually a pool of liquid mercury, and the positive ignitor is responsible for causing the gas within the tube to ionize.

Normally, two ignitrons are used together to control an AC load current. Figure 6–30 illustrates how this can be accomplished by employing two thyratrons and a phase-shifting network to control the firing delay of the ignitrons, and thus the AC load current. This circuit uses a phase-shifting network similar to that illustrated previously to shift the grid voltage of V_3 and V_4 from 0° to 180°. This controls the time (during the 180° conduction cycle) that the ignitor of each individual ignitron is positive and causes each tube to ionize. Resulting waveforms are illustrated in Figure 6–31.

In this chapter circuits that demonstrate the application of the Shockley diode, SCR, diac, triac, UJT, transistor, thyratron, and ignitron when connected in a trigger mode have been examined. Answer the following review questions, solve the problems, and perform the suggested laboratory activities to help you better understand the operation of the circuits discussed.

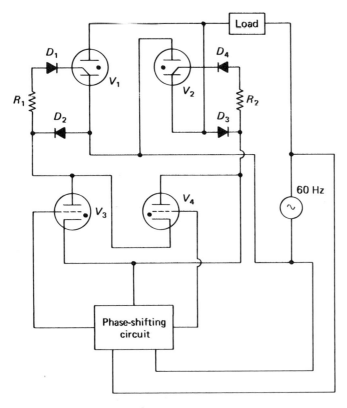

FIGURE 6–30   Typical ignitron circuit.

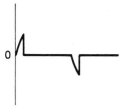

Voltage across ignitrons

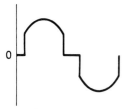

Current to load

FIGURE 6–31   Ignitron waveforms.

REVIEW

1. How is an *RL* time constant determined?

2. How is an *RC* time constant determined?

3. Discuss the *RL* time constant charging curve.

4. Discuss the *RC* time constant charging curve.

5. Discuss the *RL* time constant discharging curve.

6. Discuss the *RC* time constant discharging curve.

7. What is a universal time constant charge curve?

8. What is a universal time constant discharging curve?

9. What is a square or rectangular waveform?

10. What is a sawtooth waveform?

11. What is a differentiator circuit?

12. What is an integrator circuit?

13. What is a relaxation oscillator?

14.   What effect would decreasing the capacitance value of an *RC* saw-tooth generator have on the output?

15.   How can a Shockley diode be used in oscillator circuits?

16.   What are the differences between the conductive action of a Shockley diode used in an oscillator circuit and the same diode used in a time delay circuit?

17.   How can the Shockley diode be used to trigger an SCR?

18.   How can the SCR be used as a DC switch?

19.   What are two methods used to trigger the SCR?

20.   How do a circle and a cycle compare?

21.   An SCR has an AC voltage applied to its anode-cathode. Why can its conduction be delayed by only 90° when its gate current is a steady direct current?

22.   If the conduction of an SCR is delayed by 75° when its anode-cathode voltage is alternating current, for how many degrees will current flow through its load?

23.   How may the conduction of an SCR be delayed more than 90° when its anode-cathode voltage is alternating current?

24.   What is the maximum number of degrees that current can flow through an SCR when its anode-cathode voltage is alternating current?

25.   How can the conduction of an SCR control DC motor speed?

26.   How can a diac be used to trigger a triac?

27.   How can AC load current be controlled by a diac-triac combination?

28.   Of what importance is the *RC* time constant of R_1 and C_1 illustrated in Figure 6–9?

29.   How can a UJT be used in oscillator circuits?

30.   How can the tone produced by the oscillator in Figure 6–22 be changed?

31.   What is the relation between the *RC* time constant and the tone produced in Figure 6–22?

32.   How can a UJT be used to trigger an SCR?

33.   How can a transistor be used to trigger an SCR?

34.   What are some differences between controlling the conduction of an SCR and the conduction of a thyratron?

35. Why must the amplitude of a thyratron's AC grid voltage be greater than a voltage value that allows the thyratron to trigger when its plate voltage reaches its peak?

36. How do the thyratron and ignitron compare as control devices?

37. When are thyratrons and ignitrons used together?

PROBLEMS

1. Refer to Figure 6–1A. With the following values of R and L, calculate the RL time constant (t) of the circuits:

   a) $R = 500 \ \Omega$, $L = 200$ mH

   b) $R = 10$ kΩ, $L = 15$ H

   c) $R = 2.5$ kΩ, $L = 60$ μH

   d) $R = 200 \ \Omega$, $L = 10$ H

   e) $R = 50$ kΩ, $L = 10$ mH

2. In a series RL circuit with a maximum current of 200 mA, what are the approximate values of current through the circuit after the following time constants during its charging condition?

   a) 1.5

   b) 2.0

   c) 3.0

   d) 3.5

   e) 4.8

3. In a series RL circuit with a maximum current of 30 mA, what are the approximate values of current through the circuit after the following time constants during its discharging condition?

   a) 1.0

   b) 1.5

   c) 2.6

   d) 3.2

   e) 5.0

4. Refer to Figure 6–2A. With the following values of R and C, calculate the RC time constant (t) of the circuits:

   a) $R = 1.5$ MΩ, $C = 1.0$ μF

   b) $R = 10$ kΩ, $C = 0.002$ μF

   c) $R = 47$ kΩ, $C = 0.01$ μF

   d) $R = 50$ kΩ, $C = 10$ pF

   e) $R = 1$ kΩ, $C = 150$ μF

5. In a series RC circuit with a maximum voltage of 210 V, what are the approximate values of voltage across the capacitor after the following time constants during its charging condition?

   a) 1.5

   b) 2.0

   c) 3.0

   d) 3.5

   e) 4.8

6. In a series RC circuit with a maximum voltage of 40 V, what are the approximate values of E_C after the following time constants during its discharging condition?

   a) 1.0

   b) 1.5

   c) 2.6

   d) 3.2

   e) 4.4

7. What would be the RC time constant of R_1 and C_1 in the circuit illustrated in Figure 6–32?

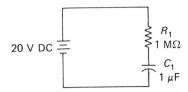

FIGURE 6–32   Time constant circuit.

8. What would be the voltage across C_1 in the circuit illustrated in problem 1 after 2 s?

9. What would be the voltage across R_1 in the circuit illustrated in problem 1 after 1 s?

10. What would be the current flow through R_1 in the circuit illustrated in problem 1 after 3 s?

SUGGESTED LABORATORY ACTIVITIES

1. RL time constant circuit.

    a) Construct an RL circuit similar to the one shown in Figure 6–1A. Use values of R, L, and E_{in} as specified by the instructor. The following are recommended: $R = 5$ Ω, $L = 10$ H, and $E_{in} = 1$ V.

    b) Calculate the RL time constant (t) in seconds.

    c) Place a DC current meter in series with the circuit.

    d) Apply voltage and observe the meter. Estimate the time required for the current to reach its maximum value.

    e) Change the resistance value and repeat steps b, c, and d.

2. RC time constant circuit.

    a) Construct a series RC circuit similar to the one shown in Figure 6–2A. Use values of R, C, and E_{in} close to those listed in the figure.

    b) Calculate the RC time constant (t) in seconds.

    c) Place a voltmeter across the capacitor.

    d) Apply voltage and observe the meter. Estimate the time required for the voltage across the capacitor to reach its maximum value.

    e) Change resistance values and repeat steps b, c, and d.

3. Differentiator circuit.

    a) Construct a differentiator circuit similar to the one shown in Figure 6–6A. Use a square-wave signal generator as the input and the values of R and C specified by the instructor.

b) Connect an oscilloscope across the output.

c) Make a sketch of the output signal.

d) Change the values of R and C several times. Observe and sketch the output signal after each component change is made.

4. Integrator circuit.

a) Construct an integrator circuit similar to the one shown in Figure 6–6B. Use a square-wave signal generator as the input and the values of R and C specified by the instructor.

b) Connect an oscilloscope across the output.

c) Make a sketch of the output signal.

d) Change the values of R and C several times. Observe and sketch the output signal after each component change is made.

5. Sawtooth generator circuit.

a) Construct a sawtooth generator circuit similar to the one shown in Figure 6–7. Use the component values specified in the figure.

b) Connect an oscilloscope across the output.

c) Make a sketch of the output signal.

d) Change the values of R and C several times. Observe and sketch the output signal after each component change is made.

6. Construct the circuit illustrated in Figure 6–33. Observe and draw the waveform across R_2.

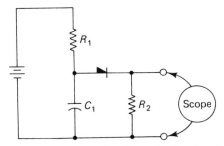

FIGURE 6–33   Shockley diode circuit.

7. Construct the circuit illustrated in Figure 6–34 and measure the time delay between the instant S_1 is closed and when the SCR is triggered.

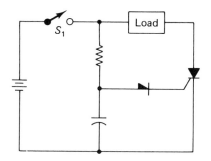

FIGURE 6–34 Shockley-SCR circuit.

8. Construct the circuit in Figure 6–35. Adjust the gate voltage to cause conduction of the SCR to be delayed by 90° as observed on the scope. Draw the resulting waveform that appears across the SCR. Observe and draw the waveform that appears across the load.

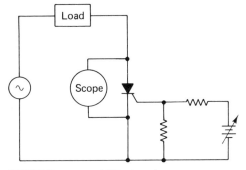

FIGURE 6–35   SCR circuit.

9. Construct the circuit in Figure 6–36. Adjust R_1 to cause a minimum conduction delay of the triac. Observe and draw the waveform that appears across the triac. Adjust R_1 to cause a maximum conduction delay of the triac. Observe and draw this waveform as it appears across the triac.

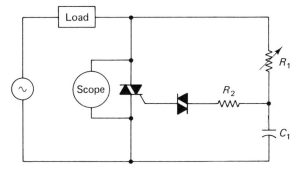

FIGURE 6–36   Diac-triac circuit.

10.   Construct the UJT oscillator in Figure 6–37. Observe and draw the waveforms across R_2.

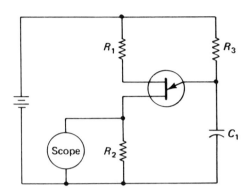

FIGURE 6–37   UJT circuit.

11.   Construct the circuit in Figure 6–38 and demonstrate how R_1 can be adjusted to turn the SCR on and off.

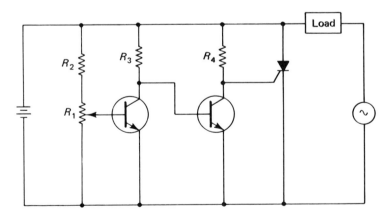

FIGURE 6–38   Transistor trigger circuit.

12.   Construct the circuit in Figure 6–39 and adjust the DC supply voltage until the thyratron's conduction is delayed by 90°. Observe and draw this waveform.

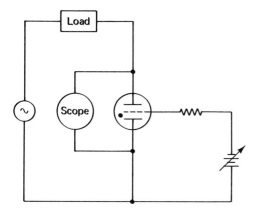

FIGURE 6–39   Thyratron circuit.

CHAPTER 7
Amplifying Devices

Amplifying devices such as NPN and PNP transistors and field-effect transistors (FETs) are most widely used in control and communications electronic circuits. Without these devices our lives would be vastly different, insomuch as long-range communication (FM, AM radio, television, etc.) would not be possible. Likewise the control of automated machines used in industry to mass produce products would not be possible without amplifying devices. Many other types of electronic equipment depend upon amplifying devices for operation.

Amplifying devices, along with appropriate circuitry, change low electrical energy levels to higher levels of electrical energy. In this chapter the conventional transistor and the FET as amplifying devices will be examined. Additionally, vacuum tube devices used as amplifiers will be briefly discussed. Even though transistors have replaced most vacuum tubes when used as amplifiers, tubes are still being used, although somewhat limitedly, in certain applications in industry.

SOLID-STATE DEVICES

CONVENTIONAL TRANSISTOR DC AMPLIFIERS

Recall from Chapter 4 that the conventional transistor may be of either the NPN or the PNP type. Additionally, the transistor has three external leads identified as emitter, base, and collector and may be connected in one of three basic configurations, the *common-base, common-emitter,* or *common-collector* configurations. Finally, the transistor exhibits two very important current gain factors: alpha (α) and beta (β).

α is an expression of the relation between the emitter and collector currents when the transistor is connected in the common-base configuration. β is an expression of the relation between the base and collector currents when the transistor is connected in the common-emitter configuration. Both α and β are often referred to as *amplification factors*.

Alpha (α)

Figure 7–1 illustrates an NPN transistor connected in a common-base configuration along with its characteristic curves. Recall the rela-

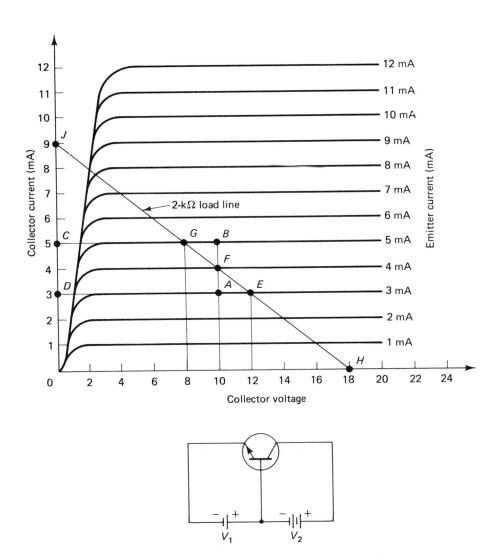

FIGURE 7–1   Transistor common-base configuration and curves.

tion between base, emitter, and collector currents from Chapter 4. As the emitter current increases, the collector current increases. The emitter current is always slightly greater than the collector current due to base leakage. The collector current plus the base current always equals the emitter current. Or, the emitter current minus the base current always equals the collector current.

The DC α of a transistor equals the collector current (I_C) divided by the emitter current (I_E) when the collector voltage (V_C) is a constant value.

$$\alpha = \frac{I_C}{I_E} \text{ with } V_C \text{ constant}$$

From the characteristic collector curves illustrated in Figure 7–1, the DC α for any given collector voltage can be computed.

The illustration shows a vertical line projected from the 10-V point on the collector voltage axis. This line intersects several emitter current curves. For purposes of discussion we are interested in intersections A and B. Intersection A is formed by the vertical projection line and the 3-mA emitter current curve. From intersection A, a horizontal projection line is extended to the left until it intersects the collector current axis at point D. This indicates that 3 mA of emitter current flow causes 2.9 mA of collector current when the collector voltage is 10 V. In this case α equals about 0.96.

$$\alpha = \frac{I_C}{I_E} = \frac{2.9 \text{ mA}}{3.0 \text{ mA}} = 0.96$$

The projection lines forming intersections at points B and C indicate 5 mA of emitter current causing 4.8 mA of collector current when the collector voltage is 10 V. Again α equals about 0.96.

$$\alpha = \frac{I_C}{I_E} = \frac{4.8 \text{ mA}}{5.0 \text{ mA}} = 0.96$$

These values reflect the static characteristics of the transistor since the currents and voltages are all steady, unchanging values.

In reality, emitter current is caused to change, resulting in a change in collector current. α is a small change in collector current (I_C) divided by a small change in emitter current (I_E) with the collector voltage (V_C) being constant (Δ means "change in"):

$$\alpha = \frac{\Delta I_C}{\Delta I_E} \text{ with } V_C \text{ constant}$$

Thus, the change in emitter current from point A to point B in the illustration may be examined. Notice that the change is equal to 2 mA ($5 - 3 = 2$ mA). This change of 2 mA in emitter current causes the change in collector current from point D to point C to equal 1.9 mA ($4.8 - 2.9 = 1.9$ mA). Thus, α equals 0.95 as a result of the changing values:

$$\alpha = \frac{\Delta I_C}{\Delta I_E} = \frac{1.9 \text{ mA}}{2.0 \text{ mA}} = 0.95$$

α is the transistor's amplification factor when this component is connected in the common-base configuration. The magnitude of output current (collector current) can be readily determined by multiplying α by the transistor's input current (emitter current) when collector voltage is constant.

Since a transistor's α is always less than unity (less than 1), the common-base configuration is used for voltage amplification rather than current amplification.

A transistor is rarely used in a circuit without additional components such as resistors or capacitors. Figure 7–2 illustrates a transistor with resistors connected in such a way as to control the transistor's conduction. R_1 is a variable resistor that controls the emitter current caused by V_1, R_L is the resistive load that, along with V_2, determines the collector voltage. Adjusting R_1 to cause a change in emitter current results in a corresponding change in collector current, a change in the voltage drop across R_L, and a change in the collector voltage. This can be demonstrated by the load line analysis illustrated in Figure 7–1.

Assume that $R_L = 2000 \ \Omega$ and $V_2 = 18$ V as illustrated in Figure 7–2. To construct the *load line* in Figure 7–1, locate reference points on the collector voltage and collector current axis. First assume that R_L is not in the circuit. This causes the collector voltage to equal 18 V for any given current. Thus, 18 V (point H) represents the collector voltage reference point.

Now assume that the transistor is not in the circuit and the current is limited only by R_L. With the transistor not in the circuit, current

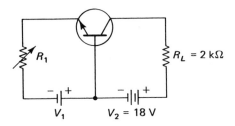

FIGURE 7–2   Transistor with resistor control.

equals 9 mA ($I = E/R = 18$ V/2000 $\Omega = 0.009$ A, or 9 mA). The collector current reference point is located at 9 mA (point J) on the collector current axis. Connecting points H and J with a very straight line forms a 2000-Ω load line. From this load line we can examine the transistor's values as the emitter current changes.

Let us assume that R_1 is adjusted to cause 3, 4, and 5 mA of emitter current to flow. These emitter currents correspond with intersection points E, F, and G formed by the load line and the emitter current curves. Thus, at point E, 3 mA of emitter current causes 2.9 mA of collector current, a collector voltage of 12 V, and 6 V across R_L. These values are determined by projecting lines from point E to the collector current and collector voltage axis and subtracting the collector voltage from source voltage V_2 (source voltage minus collector voltage equals voltage across R_L, or $18 - 12 = 6$ V).

The intersection at point F indicates an emitter current of 4 mA, which causes 3.9 mA of collector current, a collector voltage of 10 V, and a voltage across R_L of 8 V.

Finally, point G shows that when 5 mA of emitter current is caused to flow by adjusting R_1, 4.8 mA of collector current results, causing the collector voltage to equal 8 V and the voltage across R_L to equal 10 V.

Thus, when the transistor is under load (a resistor, inductor, etc., connected in series in its collector circuit), the collector voltage decreases with each increase in emitter current. Each decrease in emitter current results in a corresponding increase in collector voltage.

Beta (β)

Figure 7–3 illustrates an NPN transistor connected in the common-emitter configuration along with its characteristic curves. Recall from Chapter 4 the relationship between base, emitter, and collector currents. As base current increases, collector current increases. The collector current is always much larger than the base current. The base current plus the collector current equals the emitter current. Or the emitter current minus the base current equals the collector current.

The DC β of a transistor equals the collector current (I_C) divided by the base current (I_B) when the collector voltage (V_C) remains constant.

$$\beta = \frac{I_C}{I_B} \quad \text{with } V_C \text{ constant}$$

Using the characteristic curves illustrated in Figure 7–3, the transistor's DC β can be computed for any collector voltage. For explanation purposes, the DC β when the collector voltage (V_C) equals 4 V will be computed.

A line projected vertically from the 4-V point on the collector voltage axis intersects several base current curves. At intersection A, notice

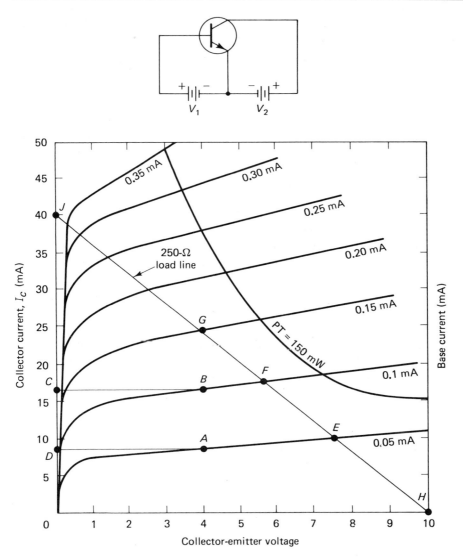

FIGURE 7–3   Common-emitter configuration with characteristic curves.

that 0.05 mA of base current causes 8 mA of collector current. This is determined by projecting a horizontal line from point A to the collector current axis (point D). In this example β computes to 160.

$$\beta = \frac{I_C}{I_B} = \frac{8 \text{ mA}}{0.05 \text{ mA}} = 160$$

Projection lines at intersection points B and C illustrate that 0.1 mA of base current results in 16 mA of collector current flow when the collector voltage equals 4 V. Again β equals 160.

$$\beta = \frac{I_C}{I_B} = \frac{16 \text{ mA}}{0.1 \text{ mA}} = 160$$

These values reflect the static state of the transistor due to all current and voltages remaining steady.

Normally, the base current of a transistor connected in the common-emitter configuration is caused to change, resulting in a changing collector current. This allows β as a small change in collector current (I_C) divided by a small change in base current (I_B) for any constant value of collector voltage (V_C) to be determined.

$$\beta = \frac{\Delta I_C}{\Delta I_B} \text{ with } V_C \text{ constant}$$

By examining the change in base current from points A to B of the illustration, and the resulting change in collector current (points D to C), the resulting β can be computed. The change in base current is found to equal 0.05 mA ($0.1 - 0.05 = 0.05$ mA). The corresponding change in collector current equals 8 mA ($16 - 8 = 8$ mA). β equals 160.

$$\beta = \frac{\Delta I_C}{\Delta I_B} = \frac{8 \text{ mA}}{0.05 \text{ mA}} = 160$$

β is the transistor's amplification factor when the device is connected in the common-emitter configuration. The output (collector current) can readily be computed by multiplying the transistor's input (base current) by β, when collector voltage is constant. A transistor's β may be as high as 400 or more. This allows the common-emitter configuration to be widely used in current amplification applications.

Like the common-base configuration, a transistor is never found in a circuit that does not contain other components. Figure 7–4 illustrates a transistor with external resistors connected to control its currents. It is important to notice that the common-emitter configuration employs only one power supply to provide bias voltage. By adjusting R_1 in the circuit, the base current may be increased or decreased. This causes a corresponding change in collector current. As collector current increases, collector voltage will decrease because of the voltage across R_L. The load line illustrated in Figure 7–3 shows these relationships.

The 250-Ω load line is plotted as previously described. Assuming R_L was not present, the transistor's collector voltage equals 10 V (point

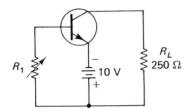

FIGURE 7–4   Common-emitter configuration with one power supply.

H). Assuming R_L to be the only limit to current flow, 10 V (source voltage) applied to 250 Ω (R_L) would result in 40 mA of collector current ($I = E/R = 10$ V/250 Ω $= 0.04$ A, or 40 mA). This allows point J to be located on the collector current axis. Points H and J are connected with a straight line.

This load line forms intersection points E, F, and G with base current curves 0.05, 0.1, and 0.15 mA, respectively. Thus, when R_1 is adjusted to cause 0.05 mA of base current, the resulting collector current equals 10 mA, the collector voltage equals 7.5 V, and the voltage across R_L equals 2.5 V. A base current of 0.1 mA allows 17 mA of collector current, which causes the collector voltage to equal 5.7 V and the voltage across R_L to equal 4.3 V.

Finally, intersection G illustrates a base current of 0.15 mA causing 24 mA of collector current, a collector voltage of 4 V, and a voltage across R_L of 6 V. These relationships illustrate that an increase in base current results in an increase in collector current. An increase in collector current causes a decrease in collector voltage when the transistor's collector is loaded (has a load connected in series).

From the previous information it becomes obvious that the transistor, with appropriate circuitry, can readily be used to cause the amplification of DC currents and voltages. Figure 7–5 illustrates a typical common-base voltage amplifier along with the transistor's characteristic curves and load line analysis. In this illustration R_1 equals 500 Ω and is used to establish the bias or emitter current flow due to V_1. By ignoring the resistance of the forward-biased emitter-base junction, which is very small, the emitter current equals 4 mA ($I = E/R = 2$ V/500 Ω $= 0.004$ A, or 4 mA). This is known as the transistor's *operating point* and is represented as point Q on the illustrated characteristic curves. Thus, 4 mA of emitter current results in 3.9 mA of collector current, a collector voltage of 12 V, and a voltage across R_L of 8 V. A negative-going 1-V input would instantaneously change the emitter-base bias voltage from 2 V (V_1) to 3 V. This additional 1 V would cause the emitter current to rise to 6 mA as represented by point P on the illustration, resulting in 5.9 mA of collector current, a collector voltage of 8 V, and a voltage across R_L of 12 V.

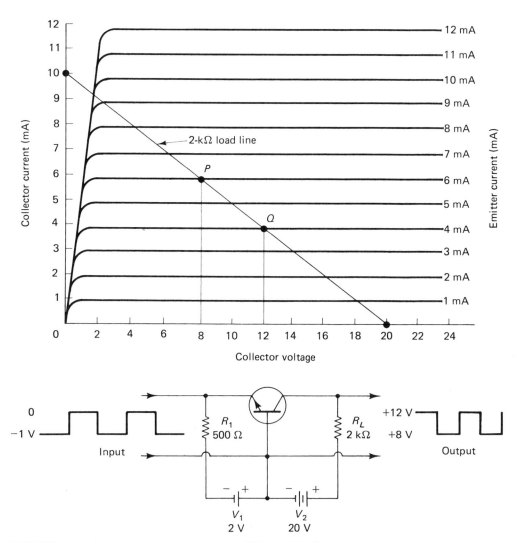

FIGURE 7–5   Common-base voltage amplifier with characteristic curves.

Thus, when the base-emitter voltage is changed 1 V, from 2 to 3 V, the collector-emitter voltage is changed 4 V, from 12 to 8 V.

   Voltage gain can be discussed as a ratio of output to input. Thus, a change of 1 V input results in a change of 4 V output providing a voltage gain of 4 V.

$$\textbf{Gain} = \frac{\Delta\textbf{output}}{\Delta\textbf{input}} = \frac{4}{1} = 4$$

As previously stated, the transistor can be used to cause current amplification. Figure 7–6 illustrates a transistor connected in the common-emitter configuration. Likewise, the transistor's characteristic curves and load line analysis are included in this figure.

Assume the value of R_1 in the illustration to equal 90 kΩ. With 9 V applied to 90 kΩ, the base current equals 0.1 mA and is represented as point Q on the illustration. This 0.1 mA of base current causes 17 mA of collector current as shown. Assume an input that would cause the base current to instantaneously rise from 0.1 to 0.2 mA (point P). This would

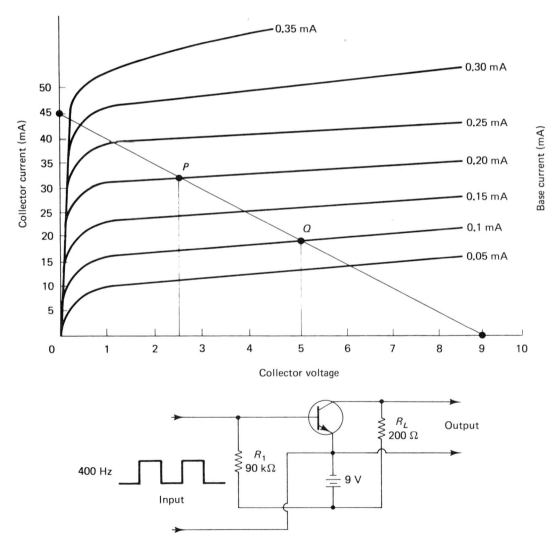

FIGURE 7–6   Transistor load line analysis.

cause an instantaneous rise in collector current from 17 to 32 mA. Thus, a change of 0.1 mA of base current $(0.2 - 0.1 = 0.1$ mA) results in a change of 15 mA of collector current $(32 - 17 = 15$ mA).

Since current gain can be described as a ratio of a change in output current to a change in input current, the illustrated amplifier demonstrates a current gain of 150.

$$\text{Gain} = \frac{\Delta\text{output}}{\Delta\text{input}} = \frac{15 \text{ mA}}{0.1 \text{ mA}} = 150$$

In this example, the output current is 150 times greater than our input current.

A variation of the transistor being used as a DC amplifier is illustrated in Figure 7–7. In this example the transistor is used in a time delay mode and controls a load current as illustrated. In this circuit, when S_1 is closed, C_1 begins charging to the source voltage through the emitter-base junction and R_1. The charging current caused by the action of C_1 is also the base current of the transistor. Thus, the charging current of C_1, or the transistor's base current, results in collector current flow through the load.

As C_1 becomes charged, the charging current and resulting base current decrease. This causes the current flow through the load (collector current) to decrease. When C_1 becomes fully charged, charging current, and thus base current, will cease to flow. With no bias current, the load current (collector current) likewise will cease to flow. Thus, when S_1 is closed, current will immediately flow through the load and will continue to flow for a certain amount of time. The amount of time that load current flows is determined by the RC time constant of R_1 and C_1. If S_2 is momentarily closed, C_1 is discharged and the cycle is repeated.

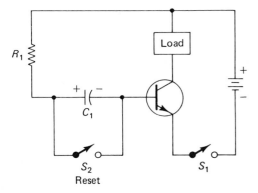

FIGURE 7–7   Time delay DC transistor amplifier.

CONVENTIONAL TRANSISTOR AC AMPLIFIERS

Transistors can cause the amplification of alternating currents and voltages just as transistors can be used as DC amplifiers. Figure 7–8 illustrates a common-base amplifier used to amplify an AC voltage. V_1 equals 3 V and causes 4 mA of input or emitter current because of R_1 ($I=$

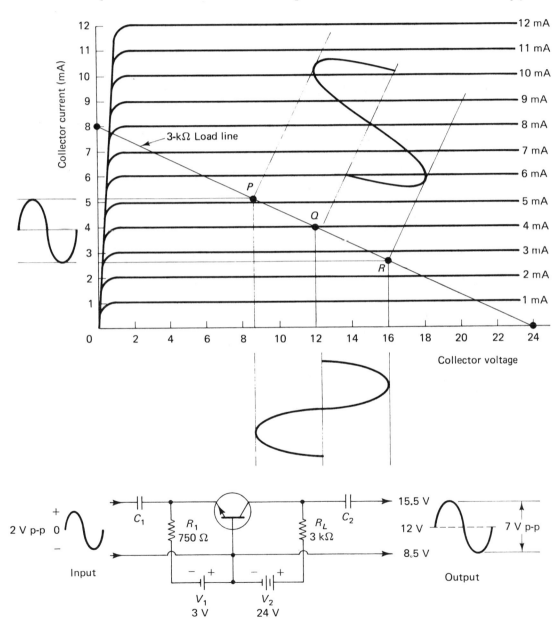

FIGURE 7–8   Transistor AC amplifier.

$E/R = 3$ V/750 $\Omega = 0.004$ A, or 4 mA). This is represented as point Q on our illustration and causes the collector voltage to equal 12 V. When an AC voltage of 2 V peak-to-peak is applied to the input, the transistor's biasing voltage and current rise and fall with the alternating current. During the positive half of the AC cycle the positive 1-V peak input is added to the negative 3 V supplied by V_1, causing the transistor's emitter-base voltage to drop to -2 V $(+1$ peak volt$) + (-3V_{DC}) = -2V_{DC}$. This instantaneous reduction in emitter-base voltage causes the emitter current to drop to about 2.6 mA (point R). This reduces the collector current and allows the collector voltage to rise to about 15.5 V.

During the negative half of the AC cycle, the negative 1-V peak input is added to the negative 3 V from V_1, causing the transistor's emitter-base voltage to rise to -4 V $(-1$ peak volt$) + (-3V_{DC}) = -4V_{DC}$. This causes the emitter current to rise to about 5.2 mA (point P). This instantaneous increase in emitter current brings about an increase in collector current and causes the collector voltage to drop to about 8.5 V. Thus, a 2-V peak-to-peak AC input voltage results in a peak-to-peak charge of 7 V ($15.5 - 8.5 = 7$ V) in the output. This results in a voltage gain of about 3.5.

$$\text{Gain} = \frac{\Delta \text{output}}{\Delta \text{input}} = \frac{7 \text{ V}}{2 \text{ V}} = 3.5$$

An important characteristic concerning the common-base AC amplifier is that its input and output voltages are in phase.

Figure 7–9 illustrates a common-emitter amplifier. Unlike the common-base amplifier, the output is 180° out of phase with the input. The positive half of the AC voltage input causes the base-emitter junction of the transistor to become more positive and thus more forward biased. This causes more base current, resulting in more collector current. This causes the collector voltage to decrease or become less positive. Thus, as the base becomes more positive, the collector becomes less positive, resulting in the input and output being 180° out of phase.

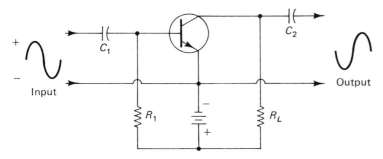

FIGURE 7–9   Amplifier input-output phase relationship.

FIELD-EFFECT TRANSISTOR (FET)

The junction field-effect transistor (JFET) is a solid-state amplifying device that is capable of producing very high gains when connected into appropriate circuitry. Figure 7–10 illustrates the electrical symbol and the physical characteristics of both the N- and P-channel FET. As shown in the illustration the basic FET has three external leads identified as the source, drain, and gate. Both the N- and P-channel FETs exhibit the same operating characteristics as the NPN and PNP transistors. The only differences relate to an opposite voltage polarity applied to their external leads, as shown in Figure 7–11. For this reason, the remainder of this explanation concerning the operation of the JFET will be devoted to the N-channel JFET.

As illustrated in Figure 7–11 the basic FET is made of a conductive channel of N material between the source and drain leads. On either

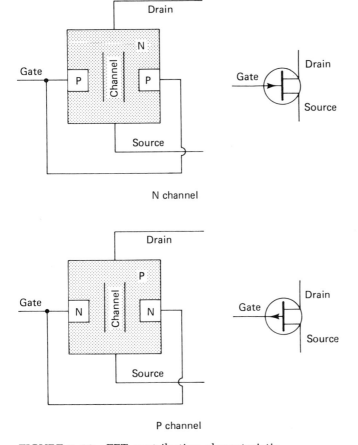

FIGURE 7–10   FET contribution characteristic.

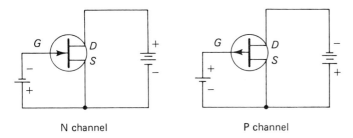

N channel   P channel

FIGURE 7–11   N and P channel FETs.

side of this channel is a small doped area of P material. Two junctions and depletion zones are formed between the P materials and the conductive channel (N material).

Recall that the physical characteristics of a conductor or conductive path determine the resistance of that conductor or path. These characteristics, among others, include

1. the length of the conductor or path,

2. the temperature of the conductor or path, and

3. the cross-sectional areas of the conductor or path.

Generally, the greater the length of the conductor, the greater its resistance. The higher the temperature of conductors that exhibit a positive temperature coefficient, the greater the resistance. The opposite is true for conductors that exhibit a negative temperature coefficient. Finally, the greater the cross-sectional area (or its diameter if it is round) of the conductor, the lower its resistance. Likewise, the smaller the cross-sectional area of the conductor, the higher its resistance. This characteristic, concerning cross-sectional areas versus a conductor's resistance, is of special interest and can be used to explain the operation of the basic FET.

Under normal operating conditions, the N-channel FET's source is negative and its drain is positive. With no voltage applied to its gate, the source-drain current (I_{SD}) that results is limited only by the resistance of the channel material and any external resistance in the source-drain circuit. This is illustrated in Figure 7–12.

If the gate is made negative with respect to the source, the two junctions formed between the gate and the channel are reverse biased. This causes the depletion zone around each junction to become wider. Since there are no current carriers in these depletion zones, the width of the conductive channel becomes more narrow. Reducing the width of the conductive channel in effect reduces its cross-sectional area and in-

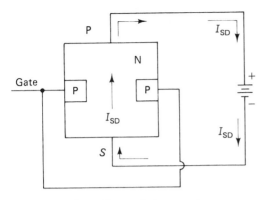

I_{SD} = Source-drain current

FIGURE 7–12   FET source-to-drain resistance.

creases its resistance. This increase in resistance reduces the source-drain (I_{SD}) current.

An increase in the reverse voltage between the gate and source further reduces the cross-sectional area of the channel. This causes a proportional increase in channel resistance and an additional reduction in source-drain current (I_{SD}). A reverse gate-source voltage great enough to stop all source-drain current is known as the FET's *"pinch-off" voltage*. Figure 7–13 illustrates several circuits with different gate-source voltages and the resulting source-drain current.

Figure 7–13 also illustrates the effect of reverse gate-source voltage. Figure 7–13A illustrates a −1 V applied to the FET's gate. This allows 5 mA of source-drain current due to the established channel width. Figure 7–13B illustrates that a −2 V on the gate reduces the width of the conductive channel to allow 3 mA of source-drain current. Figure 7–13C illustrates that a −4 V connected to the gate closes the conductive channel between the source and drain and "pinches off" the current. In this instance a −4 V equals the pinch-off voltage of the FET.

Recall that the conventional NPN or PNP transistor is a current-operated device. Input current (emitter or base current, depending upon the configuration) controls the output current (collector current). It becomes obvious that the FET is a voltage-operated device. Input voltage (gate voltage) controls output current (source-drain current). Figure 7–14 illustrates the source-drain characteristics of an FET.

The circuit in Figure 7–14A is associated with intersection points *M*, *N*, and *O* on the characteristic curves. This illustration of the *static characteristics* of the FET indicates that a constant source-drain voltage and a changing gate-source voltage bring about a changing source-drain current. Thus, if the source-drain voltage is 9 V and the gate voltage is − 3.0, − 2.5, and − 2.0 V, the source-drain current becomes 2, 3, and

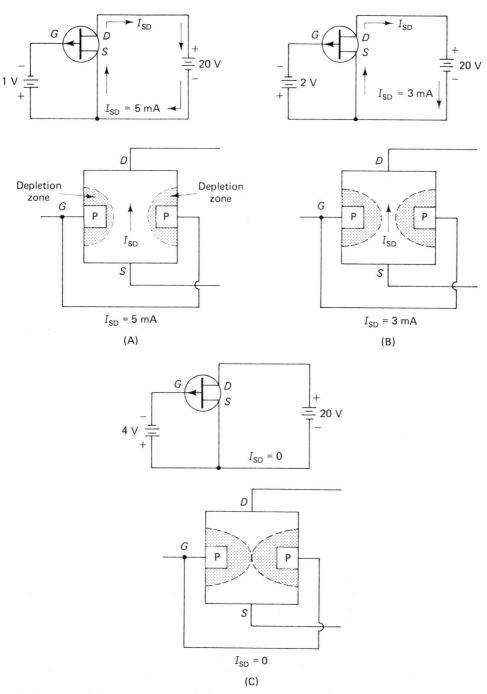

FIGURE 7–13 Gate-to-source voltage versus source-to-drain current. (A) $I_{SD} = 5$ mA. (B) $I_{SD} = 3$ mA. (C) $I_{SD} = 0$ mA.

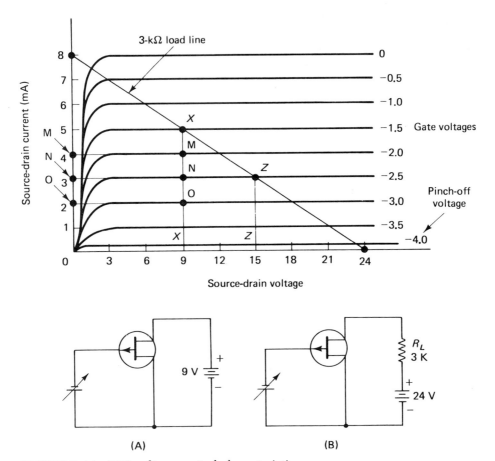

FIGURE 7–14   FET voltage control characteristics.

4 mA, respectively (points O, N, and M). As the gate becomes less negative, the source-drain current increases.

The circuit in Figure 7–14B illustrates the *dynamic characteristics*
of the FET. Load resistance R_L equals 3000 Ω and a load line is plotted
following the procedures previously described. As indicated by intersection points X and Z, a change of 1 V on the gate $(2.5 - 1.5 = 1$ V$)$ causes
a change of 2 mA of source-drain current $(5 - 3 = 2$ mA$)$ and a change in
the source-drain voltage of 6 V $(15 - 9 = 6$ V$)$.

The metal-oxide semiconductor FET *(MOSFET)* is also called an in-
sulated-gate FET *(IGFET)* and is illustrated in Figure 7–15. Notice that
this device has four external leads identified as the source, drain, gate,
and substrate. Like the JFET the MOSFET may be either a P or an N
channel. Unlike the JFET the MOSFET may be used in either a depletion
or an enhancement mode.

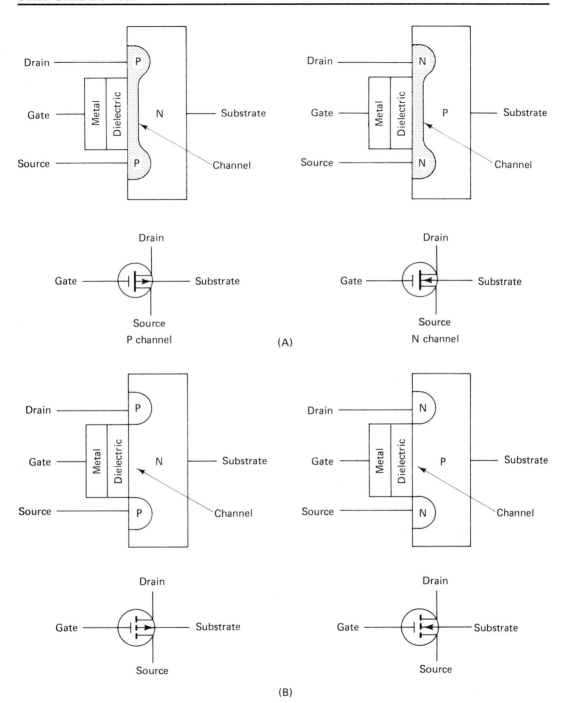

FIGURE 7–15 Metal-oxide FET (MOSFET) depletion mode
(A) and enhancement mode (B).

Depletion Mode

The *depletion-type MOSFET* does not require its gate to be biased for source-drain current to flow because of the semiconductor channel provided in the manufacturing process. The gate of the MOSFET may be either positive or negative and controls the width of the channel (and thus its resistance and source-drain current) electrostatically due to the capacitance formed between the metal gate and the corresponding substrate area. Thus, with an N-channel depletion-type MOSFET, a positive gate voltage tends to make the channel more negative and causes an increase in source-drain current. Likewise, a negative gate voltage causes the channel to become less negative resulting in a decrease in source-drain current. When the gate becomes negative enough to "deplete" the channel of current carriers, the source-drain current is at pinch off.

Enhancement Mode

The *enhancement-type MOSFET* requires its gate to be reverse biased for source-drain current to flow. This reverse biasing (gate must be positive for the MOSFET N channel and negative for the MOSFET P channel) actually provides or "enhances" the conductive channel between the source and drain. As previously described, this is brought about electrostatically due to the capacitance formed between the gate and substrate. When no reverse bias exists between the gate and the source, no conductive channel exists and no source-drain current can flow.

Both the depletion and enhancement MOSFETs exhibit a very high input impedance because of the gate-substrate capacitance. Under normal operating conditions, this capacitance will not allow input current (gate-source) to flow.

One of the most important characteristics of all FETs is its *transconductance* as measured in mhos or more commonly, micromhos. (Mhos is ohms spelled backwards and expresses how well a device conducts rather than how well it inhibits conduction.) Transconductance (G_{fs}) of an FET is an expression of the ratio of change brought about in source-drain current for any corresponding change in gate-source voltage. Thus,

$$G_{fs} = \frac{\Delta I_{SD}}{\Delta V_{GS}} \times 1,000,000$$

where:

G_{fs} is transconductance in micromhos

I_{SD} is source-drain current in DC milliamperes, and

$V_{GS} = $ DC gate-source voltage

The transconductance of FETs can range from 30 to 40 micromhos up to 40,000 or 50,000 microhmos, depending upon classification and manufacturer.

V-MOSFET Operation

The V-MOSFET shown in Figure 7–16 is a high-voltage power transistor which may be operated at high frequencies. The gate area of a V-MOSFET is V-shaped as shown in the illustration. The n^+ layers of the device are heavily doped, low-resistive material, while the n^- layers are lightly doped and higher resistant. A layer of silicon dioxide (SiO_2) covers the top horizontal surface shown in the illustration on the V area. The source terminals connect through the SiO_2 area to the n^+ and p layers of the device. The lower n^+ layer is the drain terminal connection.

The operating characteristics of a V-MOSFET are similar to those of enhancement mode MOSFETs. When the gate terminal is made more positive with respect to the source, the channel resistance decreases, thus increasing drain current flow. Increased drain current is possible due to reduced channel length. The primary advantages of the V-MOS-FET over other MOSFETs are the higher voltage and power capabilities made available by the unique construction of this device.

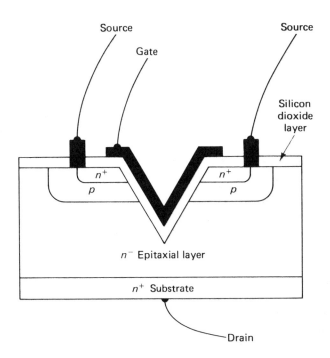

FIGURE 7–16   Cross section of N channel V-MOSFET.

Manufacturer's Specifications

It becomes obvious that the FET exhibits a wide range of electrical characteristics, and thus specifications. Some of the more common manufacturer's specifications include the maximum gate-source voltage, maximum source-drain current, maximum gate-drain voltage, gate-source avalanche voltage, gate-source pinch-off voltage, maximum power dissipation, and common-source forward transconductance, all listed relative to an operating temperature of 25 °C.

The *maximum gate-source voltage* is the largest amount of voltage that may be applied between gate and source connections. The *maximum source-drain current* indicates the largest current in milliamperes that may pass through the FET's conductive channel for any given source-drain voltage. The *maximum gate-drain voltage* represents the greatest voltage potential that can exist between the gate and drain connections. The *gate-source avalanche voltage* is an expression of reverse voltage that causes reverse current between the gate and source terminals. The *gate-source pinch-off voltage* is the gate-source voltage that closes the FET's conductive channel. Additionally, the *maximum power dissipation* is the greatest amount of power that can be converted by the FET and is usually listed in milliwatts. Finally, the *common-source forward transconductance* is an expression of how well the FET conducts when connected in the common-source configuration. This value is listed in microhmos for a specific operating frequency.

VACUUM TUBE DEVICES

As previously stated, solid-state devices have generally replaced most vacuum tubes. Because the vacuum tube is still being used in limited quantities and in specialized applications, the following space is provided to examine and explain its characteristics.

TRIODE

The *triode vacuum tube* allows current flow between its negatively charged cathode and its positively charged plate due to thermionic emission, as explained in Chapter 1. The triode tube exhibits many of the same physical and electrical characteristics of the diode vacuum tube explained in Chapter 2, with one major exception. *Tri* means three and *ode* means electrode; thus the triode vacuum tube has three active electrodes or internal elements, while the diode has only two (*di* means two). Figure 7–17 illustrates the electrical symbol for the triode vacuum tube.

From this illustration, it is seen that the triode tube possesses the

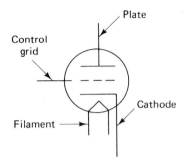

FIGURE 7–17   Triode vacuum tube.

same types of elements as does the diode. The triode tube has a filament that is connected externally to a voltage source. This external voltage source provides current flow through the filament which causes the filament to give off heat. This heat is transferred to the cathode, which is negatively charged. The cathode emits electrons that are attracted to the positively charged plate.

The significant difference between the operation of the diode vacuum tube and the triode tube is the third element, the *control grid*. This element is placed between the tube's cathode and plate and is in the path of the electrons emitted by the cathode as they are attracted toward the triode's plate. The control grid can be used to control the tube's conduction when it is appropriately biased or charged.

Normally, the grid of the triode is made negative as compared to the triode's cathode, as illustrated in Figure 7–18. The control grid is constructed of a fine-mesh screen. Since it is normally negative as compared to the tube's cathode, it repels some of the electrons that ordinarily reach the plate. The more negative the tube's control grid, the fewer electrons emitted by the cathode reach the plate. The grid can become

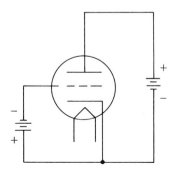

FIGURE 7–18   Element connections of triode.

negative enough to stop or cut off all electrons from reaching the positively charged plate. Thus, by controlling the negative *grid voltage* of the triode, the tube's *plate current* is controlled.

Figure 7–19 illustrates a typical *family of characteristic plate curves* for a triode vacuum tube. From this illustration it is seen that for any given plate voltage, there results a specific plate current for any given negative grid voltage. Further, the more negative the grid becomes, the less the plate current's value. This represents the tube's characteristics during its *static state* (no load in its plate-cathode circuit, all steady-state values).

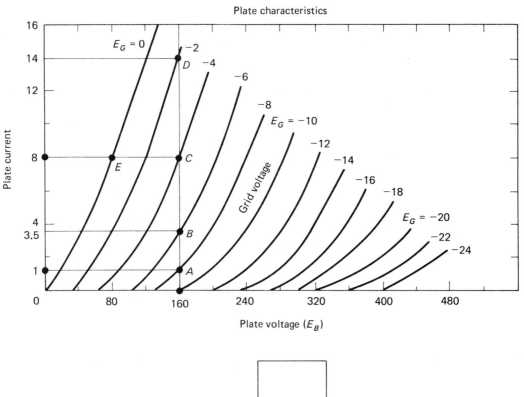

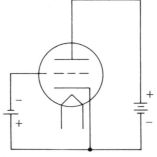

FIGURE 7–19   Family of characteristic plate curves of triode.

By examining intersection points *A, B, C,* and *D* on the graph, it can be seen that when the plate voltage remains unchanged at 160 V, and the grid voltage is −8 V, the plate current equals 1 mA (point *A*). As the grid voltage becomes less negative and is reduced to −6 V (point *B*), the plate current rises to about 3.5 mA. A further reduction in grid voltage to −4 V (point *C*) allows the plate current to rise to 8 mA. Finally, when the grid voltage is reduced to −2 V (point *D*), the resulting plate current rises to about 14 mA.

From the curves illustrated in Figure 7–19 the triode's most commonly used constants—*amplification factor* (μ), *plate resistance* (R_p), and *transconductance* (G_m)—can be determined.

The amplification factor is an expression of a change in grid voltage (ΔE_G) bringing about a change in plate voltage (ΔE_P) when plate current is constant.

$$\mu = \frac{\Delta E_P}{\Delta E_G}$$

By examining the tube's amplification factor when plate current (I_P) equals 8 mA, it can be determined that 0 grid volts (point *E*) causes plate voltage (E_P) to equal 80 V. When the grid voltage changes to become −4 V (point *C*), plate voltage (E_P) rises to 160 V. Thus, a change of 4 V on the tube's grid $(4 − 0 = 4 \text{ V})$ brings about a change in the plate voltage of 80 V $(160 − 80 = 80 \text{ V})$. The tube's amplification factor equals 20 when the plate current is constant at 8 mA.

$$\mu = \frac{\Delta E_P}{\Delta E_G} = \frac{80}{4} = 20$$

The tube's DC plate resistance (R_P), which is its internal resistance, may be computed by using Ohms law, incorporating plate voltage (E_P) and the resulting plate current (I_P) for any given grid voltage.

$$R_P = \frac{E_P}{I_P}$$

Figure 7–19 indicates that when the triode's grid voltage equals −4 V (point *C*), I_p equals 8 mA and E_p equals 160 V. Thus the plate resistance at point *C* equals 20 kΩ.

$$R_P = \frac{E_P}{I_P} = \frac{160 \text{ V}}{0.008 \text{ A}} = 20,000 \text{ Ω}$$

Obviously, if an AC voltage was applied between the grid and cathode of the tube, the tube's resistance would constantly change with the change in grid voltage. The tube's range of resistance between points on the graph may then be analyzed.

As with the FET, transconductance (G_m) is an indication of how well the triode conducts as measured in mhos or micromhos and indicates the relation between a change in grid voltage (E_G) and the resulting change in plate current (I_P) when plate voltage (E_P) is held constant.

$$G_m = \frac{\Delta I_P}{\Delta E_G} = \text{mhos}$$

By examining the change in grid voltage from points C to D on the curves, we find that at point C, -4 V on the grid causes a plate current of 8 mA. When the grid voltage changes to -2 V, a resulting plate current of 14 mA is realized. Thus, a change of 2 V on the grid (4 V at point $C - 2$ V at point $D = 2$-V change) causes a change of 6 mA of plate current (14 mA at point $D - 8$ mA at point $C = 6$-mA change). The tube's transconductance equals 0.003 mhos or 3000 micromhos.

$$G_m = \frac{\Delta I_P}{\Delta E_G} = \frac{0.006 \text{ A}}{2 \text{ V}} = 0.003 \text{ mhos} \quad \text{or 3000 micromhos}$$

Figure 7–20 represents the triode's dynamic characteristics (a load in the plate-cathode circuit).

By applying a 22.5-kΩ load line to these curves (from point P to point R) the tube's characteristics while under load can be observed. The intersection formed by the load line and the -4 grid voltage curve is indicated as point Q on the graph and represents the *operating point*. By projecting lines down to the plate voltage axis and across to the plate current axis from point Q, notice that $E_P = 140$ V and $I_P = 6$ mA. The voltage across the resistive load (R_L) equals 130 V ($270 - 140 = 130$ V). If the grid voltage becomes -2 V (intersection point M), the plate voltage becomes 110 V, the plate current becomes 7 mA, and the voltage across R_L becomes 160 V. Likewise, if the grid voltage becomes -6 V (intersection point N), the plate voltage rises to 170 V, plate current drops to 4.5 mA, and the voltage across R_L becomes 100 V.

From the preceding discussion, it becomes obvious that as the triode's grid becomes more negative, plate current decreases and plate voltage increases. Likewise, as the tube's grid voltage becomes less negative, plate current increases, while plate voltage decreases.

TETRODE

The tetrode tube is very similar to the triode tube with one major exception. The tetrode tube has an additional grid inserted between the tube's plate and control grid. This additional grid is known as the *screen*

Plate characteristics

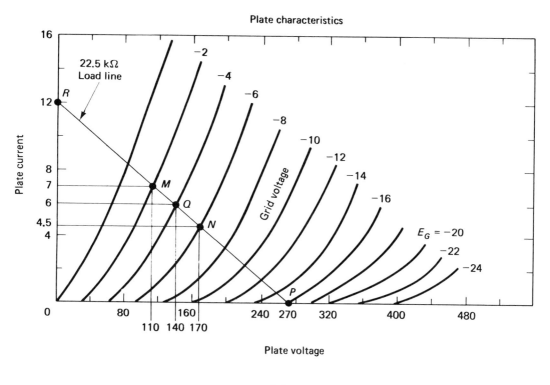

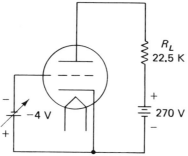

FIGURE 7–20   Dynamic characteristics of triode.

grid and is illustrated in Figure 7–21. The placement of the screen grid between the control grid and plate reduces the *interelectrode capacitance* that exists during high-frequency applications and makes it possible to achieve greater amplification.

The screen grid is usually positive as compared to the tube's cathode. Because it is so porous, most of the electrons that are emitted by the cathode pass through the control grid, are accelerated by the screen grid, and pass through to the plate.

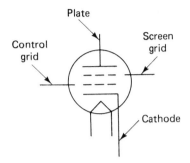

FIGURE 7–21   Elements of tetrode.

PENTODE

The pentode is similar to the tetrode tube in characteristics insomuch as it has a cathode, control grid, screen grid, and plate. The major difference between the tetrode and pentode tube is the addition of a third grid, called the *suppressor grid*, as illustrated in Figure 7–22.

The suppressor grid is inserted between the screen grid and the plate of the tube to correct a condition known as secondary emission. The accelerating effect on the electrons caused by the influence of the screen grid caused electrons to strike and "bounce" off of the tube's plate, rather than being conducted by the plate. This is known as *secondary emission* and reduces the efficiency of the tube. With the insertion of the suppressor grid, which is normally as negative as the cathode, the electrons are slowed in their travel to the plate. When secondary emission does occur, the negatively charged suppressor grid repels those electrons that are bounced off, back to the tube's plate.

The pentode tube is known for its high amplification factor of 400 or more.

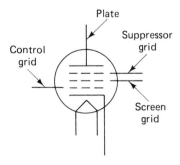

FIGURE 7–22   Elements of pentode.

In this chapter amplifying devices that include transistors, FETs, and vacuum tubes have been examined, with an opportunity provided to study DC and AC amplifiers. Answer the following review questions, solve the problems, and perform the suggested laboratory activities to gain a better understanding of the information presented in this chapter.

REVIEW

1. What is accomplished by amplifying devices and appropriate circuitry?

2. What is a transistor's alpha (α) characteristic and why is it always less than 1?

3. What is a transistor's beta (β) characteristic and why do we normally associate β with current gain?

4. What transistor configuration is normally associated with α and β?

5. Why is it more advantageous to examine a transistor's changing characteristics rather than its steady characteristics?

6. How may a transistor's β be determined?

7. Why is β always greater than α?

8. What can be determined by a transistor's load line?

9. Describe how a transistor's load line is drawn.

10. What is meant by gain and how is gain computed?

11. How may a transistor be used in time delay circuitry?

12. What is the difference between DC and AC amplification?

13. What is the difference between a transistor's input and output and how do these compare?

14. What is the phase relation between the input and output of the common-emitter amplifier? Explain why this relation exists.

15. What is the phase relation between the input and output of the common-base amplifier? Explain why this relation exists.

16. What is a junction field-effect transistor?

17. How does the JFET differ from a PNP or an NPN transistor?

18. How is the JFET's source-drain current controlled by its gate?

19. What is the difference between an N-channel and P-channel JFET?

20. What are three quantities that control the resistance of a conductive path?

21. What is a JFET's pinch-off voltage?

22. What are the differences between a JFET and a MOSFET?

23. What is the difference between the MOSFET and the IGFET?

24. What are the differences between the depletion-type MOSFET and the enhancement-type MOSFET?

25. What is transconductance and what is its unit of measurement?

26. What manufacturer's specifications are important to the operation of all FETs?

27. How does the triode tube differ from the tetrode tube?

28. How does the tetrode tube differ from the pentode tube?

29. How may a tube's amplification factor be computed?

PROBLEMS

1. Using the collector voltage curves found in Figure 7–1, compute the transistor's DC α when the collector voltage is 6 V and emitter current is 4 mA.

2. Using the collector voltage curves found in Figure 7–1, determine the collector voltage for 3 mA of emitter current if the load line is drawn as 3000 rather than 2000 Ω.

3. Using the collector voltage curves illustrated in Figure 7–3, compute the transistor's DC β when the collector voltage is 3 V and base current is 0.15 mA.

4. Using the collector curves found in Figure 7–3, determine the collector voltage for 0.1 mA of base current if the load line is drawn as 350 rather than 250 Ω.

5. Compute the gain of an amplification circuit with 0.5 V of input resulting in a change of 20 V of output.

6. Compute the transconductance for an FET with a DC gate-source voltage of 0.5 V causing 10 mA of source-drain current.

7. Using the plate voltage curves illustrated in Figure 7–19, compute the tube's amplification factor at 6 mA when the grid voltage changes from -2 to -4 V.

8. Using the plate voltage curves illustrated in Figure 7–19, compute the tube's DC plate resistance when plate voltage is 320 V and the grid voltage is − 14 V.

9. Using the plate voltage curves illustrated in Figure 7–19, compute the tube's transconductance when plate voltage is 200 V and grid voltage changes from − 8 to − 6 V.

10. Using the plate voltage curves illustrated in Figure 7–20, determine E_P, I_P, and voltage across the load resistor if the load line is drawn to represent 50 rather than 22.5 kΩ.

SUGGESTED LABORATORY ACTIVITIES

1. Construct the circuit illustrated in Figure 7–23 and measure the collector currents, collector voltages, and voltages across R_L for three different emitter currents.

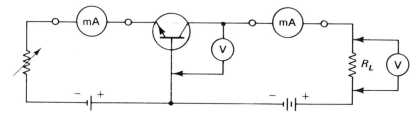

FIGURE 7–23   Common-base transistor circuit.

2. Construct the circuit illustrated in Figure 7–24 and measure the collector currents, collector voltages, and voltages across R_L for three different base currents.

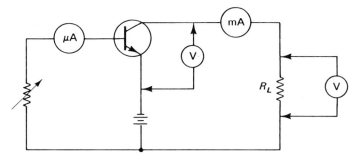

FIGURE 7–24   Common-emitter transistor circuit.

3.   Design and construct a circuit that employs a transistor, along with its *RC* control circuit, that will allow collector current to flow for 2 s.

4.   Design and construct a common-base amplifier that demonstrates AC amplification.

5.   Using the JFET, construct the circuit illustrated in Figure 7–25 and measure the source-drain current and source-drain voltage when the gate voltage is − 1.5, − 1, and − 0.5 V.

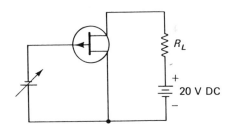

FIGURE 7–25   FET circuit.

6.   Using an appropriate reference manual, determine and record the manufacturer's specifications for three different JFETs.

7.   Using an appropriate tube manual, determine and record the manufacturer's specifications for the following vacuum tubes:

    a) Triode—6AB4

    b) Tetrode—6FV6

    c) Pentode—6AK5

8.   Using the triode, construct the circuit illustrated in Figure 7–26 and plot a family of plate characteristic curves for the vacuum tube.

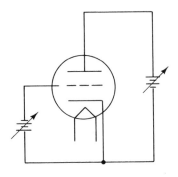

FIGURE 7–26   Triode circuit.

CHAPTER 8
Amplifier Circuit Applications

In Chapter 7 it was noted that amplification devices, with appropriate circuitry, can be classified as either *DC* or *AC amplifiers*. Amplifiers can further be classified as *audio* or *radio frequency* amplifiers, depending upon the frequency of the input signal to be amplified. Likewise, amplifying devices are used in *oscillator circuits* to aid in creating output signals at a specified range of frequencies. In this chapter the characteristics of many types of amplification circuits will be examined. Additionally, the way in which amplifier devices are biased to cause them to be classified as *class A, B, or C amplifiers* will be studied. Also, *maximum power transfer* in amplifier circuits will be discussed.

AMPLIFIER CLASSIFICATION
AND DEVICE BIASING

CLASSIFICATION

Amplifiers are classified as class A, B, or C amplifiers. *Class A amplifiers* conduct or produce an output that is identical to the amplifier's input, disregarding phase relations. That is, a class A amplifier would produce a sine wave output if a sine wave was applied as its input, as illustrated in Figure 8–1. Generally, the factors that control amplification classification are the amplitude of the input signal and the bias or operating point chosen for the amplifier.

A *class B amplifier* conducts or produces an output that represents approximately half of the amplifier's input. The amplification device used in a class B amplification circuit is biased at its "cutoff" point. Simply stated, the amplifier does not conduct until an input signal is

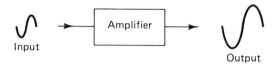

Input
Output

FIGURE 8–1   Class A amplifier.

present. Figure 8–2 illustrates the input and output of a class B amplifier.

A *class C amplifier* conducts or produces an output that represents less than half of the circuit's input. Here, the amplification device is biased below its cutoff point. That is to say, its input circuit is reverse biased. The input signal must provide the necessary electric energy to overcome the reverse-biased state of the device. This causes it to become forward biased, thus forcing its conduction. Figure 8–3 illustrates the input-output characteristics of a class C amplification circuit.

The simplest way to bias a class A amplifier is by connecting a resistor in series with its forward-bias voltage source and its input terminals. This limits input current and establishes output current and voltage. Figure 8–4 illustrates a *family of collector-emitter characteristic curves* for a common-emitter amplifier. Circuit A of Figure 8–4 illustrates an example of a class A amplifier while circuit B shows an example of class B amplification. Circuit C represents an amplification circuit that is classified as a class C amplifier. In all instances, amplification classification is determined by a combination of input terminal biasing (base-emitter) and the amplitude of the input signal.

Assume in circuit A of Figure 8–4 that R_1 is of a value that allows 0.1 mA of base current flow. From the characteristic curves illustrated in Figure 8–4, it is seen through load line analysis that 0.1 mA of base current causes collector current to equal approximately 18 mA and collector voltage to equal 5.5 V. These values reflect the steady state of the amplifier with no signal applied to its input. When a 2-V peak-to-peak (p-p) input signal is applied between the base-emitter terminals, the forward-bias voltage becomes 2.5 V during the positive half of the input signal $(+1$ peak volt$)+(+1.5V_{DC})=2.5$ V. The same bias voltage becomes 0.5 V during the negative half of the input signal $(-1$ peak volt$)+(+1.5V_{DC})=0.5$ V. Assuming a value of 15 kΩ for R_1 results in the input

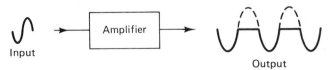

Input
Output

FIGURE 8–2   Class B amplifier.

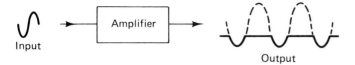

FIGURE 8–3   Class C amplifier.

current (base current) to range from approximately 0.15 to 0.03 mA. This causes collector voltage to change from 3.8 to 7.5 V and collector current to change from 26 to 6 mA. The transistor provides a complete sine wave output in this example.

In circuit B of Figure 8–4 it becomes obvious that there is no difference of potential between the transistor's base and emitter. In this instance, with no input signal, base current equals zero, collector current equals zero, and collector voltage equals 9 V. Assume a 2-V peak-to-peak signal is applied to the transistor's input (base-emitter). The negative half of the input signal (-1 V peak) reverse biases the base-emitter junction resulting in zero base current, zero collector current, and a collector voltage of 9 V. When the positive half of the input signal ($+1$ V peak) is applied to the base, the base-emitter junction is forward biased. Assume that a $+1$ V peak signal results in base current rising from 0 to 0.1 mA. As indicated by the curves illustrated in Figure 8–4, this rise in base current causes the collector current to increase from 0 to about 18 mA and collector voltage to decrease from 9 V to about 5.5 V. The transistor is conductive only during the positive half of the sine wave input and produces only half of a sine wave as its output signal.

In circuit C of Figure 8–4 the base-emitter junction is reverse biased by the 0.5-V source. Assume a 2-V peak-to-peak input signal is connected to the circuit's input (base-emitter terminals). When the negative half of the signal is applied to the base, the reverse-bias voltage becomes -1.5 V (-1 V peak) + ($-0.5V_{DC}$) = -1.5 V. The base current equals zero, the collector current equals zero, and the collector voltage equals 9 V. These values are indicated on the family of curves illustrated in Figure 8–4 and are the same as the transistor's cutoff values. In this instance the transistor is biased below cutoff.

When the positive half of the signal is applied to the base, the transistor's input is forward biased by $+0.5$ V ($+1$ V peak) + ($-0.5V_{DC}$) = $+0.5$ V. Assume this condition causes base current to rise from 0 to 0.05 mA. As shown on the transistor curves, the resulting collector current is 10 mA and the collector voltage drops to 7 V. In this example the transistor conducts for much less than half of an input sine wave signal.

BIASING

There are many approaches and considerations associated with amplification device biasing. Basically, biasing methods can be classified as

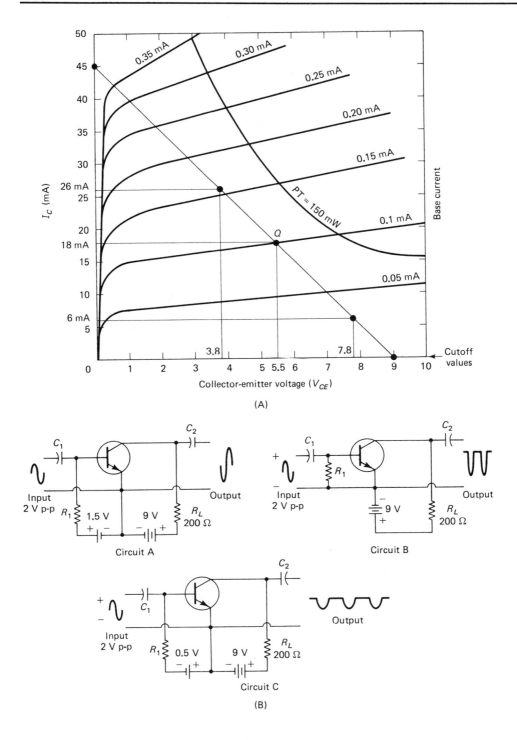

(A)

Circuit A

Input
2 V p-p

Circuit B

Input
2 V p-p

Output

Input
2 V p-p

Circuit C

Output

(B)

FIGURE 8–4   (A) Family of characteristic curves. (B) Amplification circuits.

fixed or *self-biasing*. Fixed-biasing is no more than one or more resistors connected in series or combination to the amplification device's terminals and its bias voltage source. Examples of fixed biasing have been shown frequently in previous sections and are illustrated in Figure 8–5.

Circuit A of Figure 8–5 illustrates a common-base amplifier employing one power supply. The forward-biased state of the transistor is established by the voltage divider formed by R_2 and R_3 (voltage drop across R_2 controls the positive potential at the transistor's base). The positive base potential forward biases the base-emitter junction. This causes emitter current limited by R_1. Emitter current (input) causes collector current (output).

In circuit B a common-emitter amplifier is illustrated using only one voltage source. The one voltage forward biases the base-emitter junction as well as reverse biases the collector terminal. R_1 is usually many thousands of ohms and establishes base current, which in turn establishes collector current.

One method of establishing self-biasing involves feedback. *Feedback* is defined as output energy (current/voltage from the output circuit) being fed back into the input portion of the circuit. There are two classifications of feedback. Feedback may be classified as negative (degenerative) or positive (regenerative). *Negative feedback* involves feeding output energy into the circuit's input energy when the two are 180° out of phase. Whereas *positive feedback* deals with feeding output energy into a circuit's input energy when the two are in phase. Figure 8–6 illustrates three circuits utilizing variations of negative feedback networks for self-biasing purposes.

In circuit A of Figure 8–6 collector voltage controls base bias and results in a very stable output (collector) current. Collector voltage is determined by collector current as discussed previously. Recall that as col-

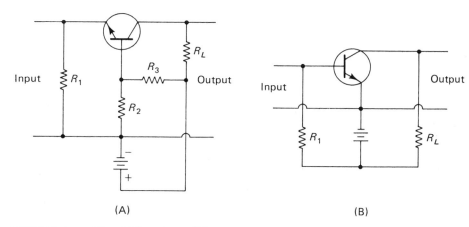

(A)   (B)

FIGURE 8–5   Fixed-biased amplifiers.

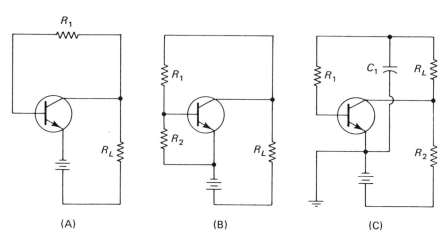

FIGURE 8–6   Self-biased amplifiers.

lector current increases, collector voltage decreases and vice versa. Thus, if collector current is caused to increase because of age or temperature, collector voltage decreases. This makes the base less positive as compared to the emitter, resulting in less base current flow through R_1. As base current decreases, a corresponding decrease in collector current results. Thus, collector current is restored to its original level and stability is maintained.

Circuit B of Figure 8–6 utilizes a combination negative-feedback, voltage divider network to provide transistor *self-biasing*. The forward biased state of the base is determined by the current flow through and the resulting voltage drop across R_1 and R_2. The current flow through R_1 and R_2 is determined by the positive state of the collector. A decrease in collector current, for any reason, causes the collector to become more positive. This increases the current flow through R_1 and R_2, resulting in an increased voltage drop across R_1 and R_2. This makes the base more positive, causing more base current, which increases collector current to its original level.

Circuit C is a variation of circuit A in Figure 8–6 and is used when an AC signal is applied to the circuit's input. Recall that the input of the common-emitter amplifier is 180° out of phase with its output. Thus, if an alteration in circuit A is not made when a signal is applied, the collector voltage variation cancels the base voltage variation resulting in no amplification. R_1 and R_2, along with the collector voltage, determine base current. R_1 is larger than R_2 and opposes AC signal current flow to the collector of the circuit. C_1 provides a path of low opposition for signal current from the collector to flow to ground. This prevents the AC output signal from being fed back to the base terminal of the circuit.

A much used self-biasing network appears in Figure 8–7. In this circuit R_2 is connected in series to the emitter and is sometimes referred to as the emitter resistor. Emitter-collector current flow develops a voltage drop across R_2, causing the emitter to become less negative. When the emitter is less negative, the base-emitter junction is less forward biased. If a change in the transistor's environment results in an increase in emitter-collector current, that increase causes the voltage across R_2 to increase. This increase in voltage drop across R_2 makes the emitter even less negative and decreases the forward-biased state of the base-emitter junction. The net result causes the emitter-collector current to return to its original level.

If an AC signal is applied to the circuit's input, C_1 is connected in parallel to R_2 and provides a path of low opposition to the signal current. This allows the signal current to "go around," or "bypass," R_2 and prevents the forward-biased state of the base-emitter junction from "floating," or increasing and decreasing, due to the signal.

AUDIO AND RADIO FREQUENCY AMPLIFIERS

Signals that cause vibrations that can be consciously heard by human beings generally range in frequencies from 20 to 20,000 Hz. Signals at frequencies that fall within this range are generally classified as audio signals, and their frequencies are classified as *audio frequencies (AF)*. Amplification circuits that are designed to amplify audio frequencies are generally classified as *AF amplifiers*.

Signals at frequencies greater than 20,000 Hz are usually classified as radio signals, and their frequencies are said to be *radio frequencies*

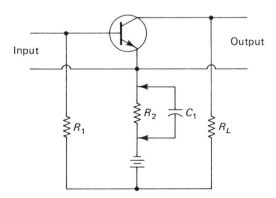

FIGURE 8–7 Emitter-biased amplifier.

(RF). Amplification circuits that are constructed to amplify radio frequencies are usually classifed as *RF amplifiers*.

The values of capacitors and inductors chosen for use in either an AF amplifier or an RF amplifier are critical due to capacitive reactance (X_C) and inductive reactance (X_L). Recall that capacitive reactance is the opposition caused by capacitance to the flow of AC currents. As the frequency of AC currents increases, capacitive reactance decreases. Capacitive reactance increases as the frequency of AC currents decreases. Likewise, recall that inductive reactance is the opposition caused by inductance to the flow of AC currents. Inductive reactance increases as the frequency of AC current increases. A decrease in the frequency of AC currents results in a decrease in inductive reactance. Usually, the values of capacitors and inductors are chosen relative to how they will react to a specific frequency or band of frequencies.

Two or more amplification circuits are commonly connected together to increase overall signal gain. When this is the case, the method used to connect, or couple, the output of one amplifier to the input of another becomes critical. Two of the most common methods of coupling amplification circuits are *capacitive coupling* and *transformer coupling*. Examples of these coupling methods appear in Figure 8–8.

When coupling amplifiers, it is extremely important to prevent the DC component of the output of the first amplification circuit from interfering with the DC bias of the input of the second amplification circuit. This is generally achieved through capacitive coupling, as illustrated in Figure 8–8A, C_1, C_2, and C_3 are *coupling capacitors*. C_1 couples the input signal to the base of Q_1. C_2 couples the collector of Q_1 (Q_1's output) to the base of Q_2. C_3 couples the collector of Q_2 (Q_2's output) to the next circuit. In all instances the values of the coupling capacitors are carefully chosen in order to cause minimum opposition to an alternating current of signal frequency. Since capacitors offer an infinite opposition to the flow of direct currents, the DC component of one circuit cannot interfere with the next circuit.

Transformers are also used to couple one amplification circuit to another, as illustrated in Figure 8–8B. The collector current of Q_1 (Q_1's output) flows through the primary of T_1. This induces a voltage across the secondary of T_1 which is coupled to Q_2's base through C_3. T_2 couples Q_2's output to the next circuit.

There are some advantages to transformer coupling in amplification circuits. First, there is no transformer action that results from direct current. Therefore, the DC component from the output of one circuit is blocked from the input of the second circuit. Second, the transformer used to couple one circuit to another may be a *step-up transformer* (more turns of wire on the secondary than on the primary) which would give greater inputs to the second amplification circuit.

Finally, transformers may also be used in coupling because of their

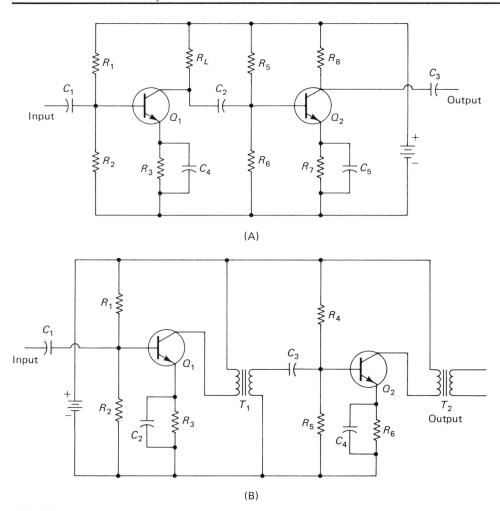

(A)

(B)

FIGURE 8–8   Capacitive and transformer coupling.

impedance-matching capabilities. A transformer can be used to cause the output impedance of one amplification circuit to match more closely the input impedance of another. This results in maximum energy transfer from one circuit to another.

MAXIMUM POWER TRANSFER

Maximum power is transferred from one circuit to another or from a voltage source to a load when the load impedance (Z_L) is equal to the source impedance (Z_S). The source impedance limits the amount of power that can be applied to a load or another circuit. A simple example is a flashlight battery. As a flashlight battery gets older, its internal resistance increases. This increase in internal resistance causes the battery to supply less power to the lamp bulb. Thus, the light output of the flashlight is reduced since less power is being transferred to the load.

Figure 8–9 shows an example which illustrates *maximum power*

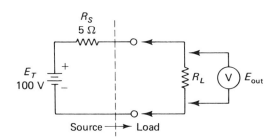

R_L (Ω)	I_L	E_{out}	Power output
0	$\dfrac{100 \text{ V}}{5 \ \Omega} = 20 \text{ A}$	20 A × 0 Ω = 0 V	20 A × 0 V = 0 W
2.5	$\dfrac{100 \text{ V}}{7.5 \ \Omega} = 13.3 \text{ A}$	13.3 A × 2.5 Ω = 33.3 V	13.3 A × 33.3 V = 445 W
5	$\dfrac{100 \text{ V}}{10 \ \Omega} = 10 \text{ A}$	10 A × 5 Ω = 50 V	10 A × 50 V = 500 W
7.5	$\dfrac{100 \text{ V}}{12.5 \ \Omega} = 8 \text{ A}$	8 A × 12.5 Ω = 60 V	8 A × 60 V = 480 W
10	$\dfrac{100 \text{ V}}{15 \ \Omega} = 6.7 \text{ A}$	6.7 A × 10 Ω = 67 V	6.7 A × 67 V = 444 W

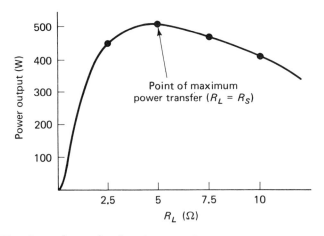

FIGURE 8–9 Circuit and graph showing maximum power transfer.

transfer. The source is a 100-V battery with an internal resistance of 5 Ω. The values of I_L, E_{out}, and power output (P_{out}) are calculated as follows:

$$I_L = \frac{E_T}{R_S} + R_L \quad E_{out} = I_L \times R_L \quad P_{out} = I_L \times E_{out}$$

Notice the graph shown in Figure 8–9 which shows that maximum power is transferred from the source to the load when $R_L = R_S$. This is an important design consideration for power supply circuits, amplifier circuits, microphones, or practically any type of electronic circuit.

REVIEW

1. What is a class A amplifier?

2. What is a class B amplifier?

3. What is a class C amplifier?

4. How can a class A amplifier be changed to become a class B amplifier?

5. What methods can be employed to achieve self-biasing of transistor amplifiers?

6. What is meant by maximum power transfer?

7. Find the values of load current, voltage output (E_{out}), and power output (P_{out}) for the circuit of Figure 8–10 using load resistance values of 0, 1, 2, 3, 4, 5, 6, 7, and 8 Ω.

8. Draw a power transfer curve using the values obtained for Figure 8–10. Plot power output (in watts) on the vertical axis and load resistance (R_L) on the horizontal axis.

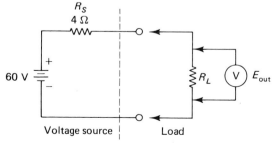

FIGURE 8–10   Maximum power transfer.

SUGGESTED LABORATORY ACTIVITIES

1. Design and construct a class A amplifier.

2. Design and construct a class B amplifier.

3. Design and construct a class C amplifier.

4. Maximum power transfer.

   a) Construct a circuit similar to the one shown in Figure 8--9. Use a fixed resistance for R_X, values of E_T, and several load resistance values specified by the instructor.

   b) With a meter, measure load current (I_L) and output voltage (E_{out}) for each value of load resistance used in the circuit. Record this data on a chart.

   c) Calculate power output (P_{out}) for each value of load resistance—$P_{out} = I_L \times E_{out}$.

   d) Plot a graph of the power transfer curve. Use the vertical axis to plot power output (in watts) and the horizontal axis to plot R_L.

CHAPTER 9
Oscillators and Frequency-Sensitive Circuit Applications

In Chapter 6 the Shockley diode and the unijunction transistor (UJT) oscillator circuit basics were examined in very simple oscillator circuits. Recall that an *oscillator* is an electronic circuit designed to create electrical waveforms or pulses at a rate of a few cycles, or pulses, per second, up to many millions of cycles per second.

An oscillator may be classified by the shape of the waveform that it produces. An oscillator that produces an output waveform shaped like a sine wave is said to be a *sinusoidal oscillator*. An oscillator that produces an output waveform that is shaped differently from a sine wave is classified as a *nonsinusoidal oscillator*. Figure 9–1 illustrates several different waveform shapes readily associated with oscillators.

SINUSOIDAL OSCILLATORS

Oscillators that produce a sine-wave-shaped output are commonly used in communication circuitry (radio, television, etc.) and depend largely upon the action of an inductor and capacitor connected in parallel to each other, commonly called a *tank circuit*.

Recall that an AC signal at resonant frequency results in X_L, caused by inductance, to equal X_C, caused by capacitance. Likewise, when an inductor and capacitor are connected in the tank circuit configuration,

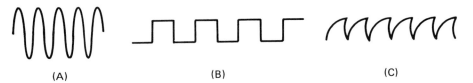

(A)   (B)   (C)

FIGURE 9–1   Oscillator waveforms.

as illustrated in Figure 9–2, the resulting waveform that is produced is at resonant frequency.

When S_1 of Figure 9–2 is closed momentarily, a pulse of direct current is caused to flow through L_1 that induces a voltage across L_1 and partially changes C_1. When S_1 is opened, the voltage induced across L_1 decays as it provides current to change C_1 to a higher voltage level. When the voltage across L_1 is smaller than the voltage to which C_1 is charged, C_1 discharges through L_1. The discharge current provided by C_1 again induces a voltage across L_1, but of opposite polarity as compared to the first polarity. When the voltage across C_1 has decayed to a level smaller than the voltage induced across L_1, it is charged again by the induced voltage across L_1, but to an opposite polarity as compared to its first polarity.

The brief pulse of direct current supplied when S_1 is momentarily closed causes current in the tank circuit to *oscillate* (flow first in one direction and then the other, or move back and forth). This is known as the *"flywheel effect"* and produces a waveform across the tank circuit as illustrated in Figure 9–2. The output waveform is described as a *"damped" waveform* because it decreases in amplitude until it eventually reaches zero. This is due to the resistance of the tank circuit. If a pulse of energy at the proper frequency is supplied to the tank circuit to overcome the effects of resistance, the output waveform is continuous (of constant amplitude) and exists as long as all remains unchanged. The tank circuit, along with appropriate amplifying circuitry, is called an oscillator circuit.

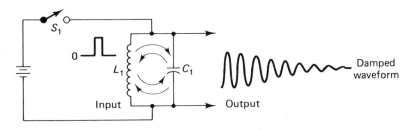

FIGURE 9–2   Tank circuit.

There are many types of oscillators that produce sine waves. The *Hartley oscillator* is a classic example and is illustrated in Figure 9–3. The tank circuit of this oscillator employs a *tapped inductor* and a *variable capacitor*. By adjusting the variable capacitor, the frequency of the oscillator's output is changed, within a given range. The transistor is biased as a class C amplifier and provides the electric energy to the tank circuit that overcomes the effects of resistance.

Transistor biasing is maintained by R_1 and R_2. C_2 couples the alternating current from the tank circuit to the transistor's base. C_4 couples the transistor's output (collector) to the tank circuit and provides the path for the feedback current that causes the tank circuit to maintain oscillation. L_3 blocks the high-frequency alternating current from the power supply.

The transistor's output is 180° out of phase with its input. Therefore, the feedback provided by the path through C_4 is 180° out of phase with the transistor's base-emitter input. This does not interfere with the transistor's conduction. Since the emitter is connected to the tap of the tank circuit's inductor and the voltage across L_1 is 180° out of phase with the voltage across L_2, the correct phase relation of the transistor's input is maintained.

Figure 9–4 illustrates a *Colpitts oscillator*. The Colpitts oscillator is very similar to the Hartley oscillator. Whereas the Hartley oscillator employed a tapped inductor in its tank circuit, the Colpitts oscillator employes a *tapped capacitor* (two variable capacitors connected in series). The feedback that maintains oscillation in the tank circuit is provided through C_5, C_2, and C_3 from a voltage divider that provides proper input signal phase between the base and emitter. The ratio between C_2 and C_3 controls the amount of feedback.

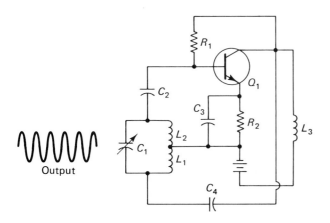

FIGURE 9–3   Hartley oscillator.

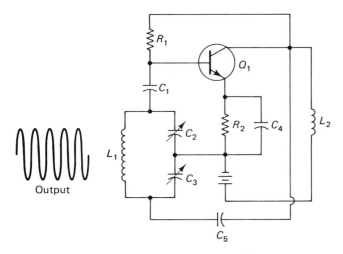

FIGURE 9–4   Colpitts oscillator.

NONSINUSOIDAL OSCILLATORS

Oscillators that produce output waveforms that are square or rectangular are called *multivibrators*. Figure 9–5 illustrates an *astable multivibrator* (astable because it constantly changes states). In this circuit the transistors are carefully matched and exhibit similar characteristics. C_1 and C_2 are the same value. Likewise, R_1 and R_2 are the same value. Finally, R_3 and R_4 are the same value. The frequency, or number of square waves per second, is determined by the RC time constants of C_1 and R_1, and C_2 and R_2.

Assume at this point that because of natural differences in the two transistors, Q_1 is off and Q_2 is on. With Q_2 conducting, C_2 charges

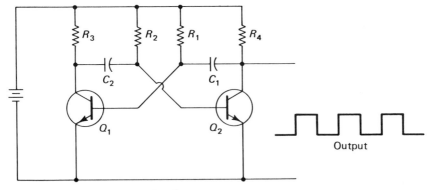

FIGURE 9–5   Astable multivibrator.

through the emitter-base junction of Q_2 and through R_3, as illustrated in Figure 9–6. As C_2 charges to the polarity indicated in the illustration, the emitter-collector current of Q_2 slowly charges C_1 through R_1 to the indicated polarity. When C_2 is charged, Q_2 stops conducting because it is no longer forward biased (voltage across C_2 very nearly equals the source voltage making the base of Q_2 as negative as its emitter). By this time C_1 has charged enough to cause the base of Q_1 to be positive and Q_1 is turned on. Since Q_2 is now off, C_1 very quickly discharges through R_4 and begins charging to the opposite polarity through the emitter-base junction of Q_1 and R_4. This is illustrated in Figure 9–7.

As C_1 of Figure 9–7 quickly discharges through R_4 and begins charging to the opposite polarity, C_2 slowly discharges through R_2. When C_2 has discharged, the collector current from Q_2 charges it to the voltage polarity indicated in Figure 9–7. When C_2 is positive enough and C_1 is negative enough, Q_2 is turned off and Q_1 is turned on. The entire cycle is repeated. Figure 9–8 illustrates the output waveform relative to the conduction of Q_1 and Q_2.

The oscillators described thus far operate continually once power is applied to the circuit and are classified as *"free-running"* oscillators. Some oscillators are not free-running and must be turned on by an external input. These types of oscillators are classified as *triggered oscillators.*

Figure 9–9 illustrates a *bistable multivibrator,* sometimes simply called a *flip-flop.* This oscillator is stable insomuch as once one of the transistors is turned on, it will remain on until triggered off by an external input. Turning off one transistor will cause the other to be turned on, thus the term *flip-flop.* In this circuit the transistors are carefully matched, the values of R_1 and R_2 are the same, the values of R_3 and R_4 are the same, and the values of R_5 and R_6 are the same.

Notice in Figure 9–9 that a small reverse voltage is connected to the bases of Q_1 and Q_2 through R_5 and R_6. This voltage alone is not great

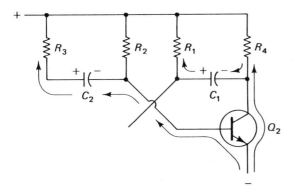

FIGURE 9–6 Astable multivibrator operation.

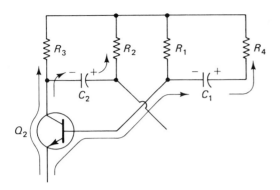

FIGURE 9–7   Astable multivibrator operation.

enough to reverse bias either base-emitter junction but can aid in keeping one of the transistors from conducting once it has been turned off.

Assume that Q_1 is on and Q_2 is off. When Q_1 is on, a large voltage
drop is developed across R_3 due to Q_1's collector current. This causes
the collector voltage of Q_1 to become very small and allows voltage V_1
through R_5 to reverse bias Q_2. This causes Q_2 to remain off. When Q_2 is
off, its collector voltage equals V_2. This makes Q_1's base even more forward biased and drives Q_1 to saturation.

A strong negative pulse applied at input 2 that overcomes the positive potential at the base of Q_1 turns Q_1 off. This causes Q_1's collector
voltage to equal V_2 and forward biases Q_2, driving it to saturation. This
causes Q_2's collector voltage to drop to almost zero and allows V_1 to reverse bias Q_1, causing Q_1 to remain off. A very large positive pulse applied to input 1 has the same effect.

FREQUENCY-SENSITIVE
CIRCUITS

Frequency-sensitive circuits are designed to respond in some way to
specific or variable AC frequencies. Some circuits are designed to pass
certain frequencies from input to output and block other frequencies.

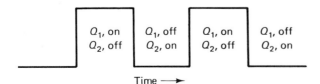

FIGURE 9–8   Astable multivibrator output.

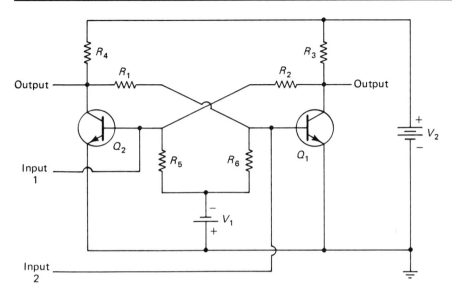

FIGURE 9–9   Bistable multivibrator.

Two types of frequency-sensitive circuits are *filter circuits* and *resonant circuits*. These circuits use reactive devices to respond to different AC frequencies and have a specific frequency response curve associated with their operation. Frequency is graphed on the horizontal axis and voltage output on the vertical axis. Frequency response curves for each type of filter and resonant circuit are discussed in the sections which follow.

FILTER CIRCUITS

The three basic types of filter circuits are shown in Figure 9–10. Filter circuits are used to separate one range of frequencies from another. *Low-pass filters* pass low AC frequencies and block higher frequencies. *High-pass filters* pass high frequencies and block lower frequencies. *Bandpass filters* pass a midrange of frequencies and block lower and higher frequencies. Filter circuits have resistance and capacitance or inductance. The reactance of capacitors or inductors makes possible the frequency selection characteristic of filter circuits.

Figure 9–11 shows the circuits used for low-pass filters and their frequency response curve. Many low-pass filters are series RC circuits (see Figure 9–11A). Output voltage (E_{out}) is taken across a capacitor. As frequency increases, capacitive reactance (X_C) decreases, since $X_C = 1/2\pi \times f \times C$. The voltage drop across the output is equal to I times X_C. So, as frequency increases, X_C decreases and voltage output decreases. Series RL circuits (see Figure 9–11B) may also be used as low-pass filters. As frequency increases, inductive reactance (X_L) increases, since $X_L = 2\pi \times$

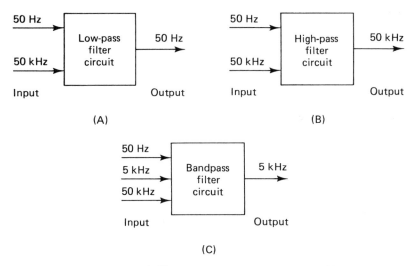

FIGURE 9–10 Types of filter circuits. (A) Low-pass filter—passes low frequencies and blocks high frequencies. (B) High-pass filter—passes high frequencies and blocks low frequencies. (C) Bandpass filter—passes a midrange of frequencies and blocks high and low frequencies.

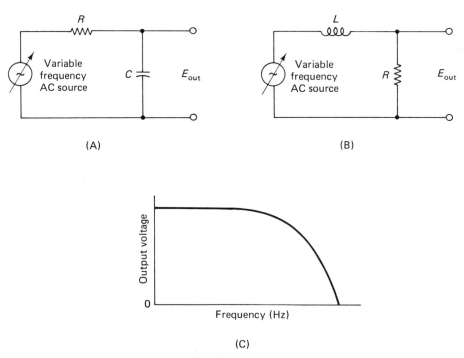

FIGURE 9–11 Low-pass filters. (A) Series RC low-pass filter. (B) Series RL low-pass filter. (C) Frequency response curve of low-pass filter.

$f \times L$. Any increase in X_L reduces the circuit's current. The voltage output taken across the resistor is equal to $I \times R$. So, when I decreases, E_{out} also decreases. As frequency increases, X_L increases, I decreases, and E_{out} decreases. Notice the shape of the frequency response curve of Figure 9–11C, showing that as frequency increases, voltage output decreases.

Figure 9–12 shows two types of high-pass filters and their frequency response curves. The series RC circuit of Figure 9–12A is a common type. The voltage output (E_{out}) is taken across the resistor (R). As frequency increases, X_C decreases. A decrease in X_C causes current flow to increase. The voltage output across the resistor (E_{out}) is equal to $I \times R$. So, as I increases, E_{out} increases. As frequency increases, X_C decreases, I increases, and E_{out} increases. A series RL circuit (see Figure 9–12B) may also be used as a high-pass filter. In this circuit the E_{out} is taken across the inductor. As the applied frequency increases, X_L increases, E_{out} also increases. In this circuit, as frequency increases, X_L increases and E_{out} increases. Notice the shape of the frequency response curve of Figure 9–12C, showing that as frequency increases, voltage output increases.

The bandpass filter of Figure 9–13A is a combination of low-pass and high-pass filter sections. It is designed to pass a midrange of fre-

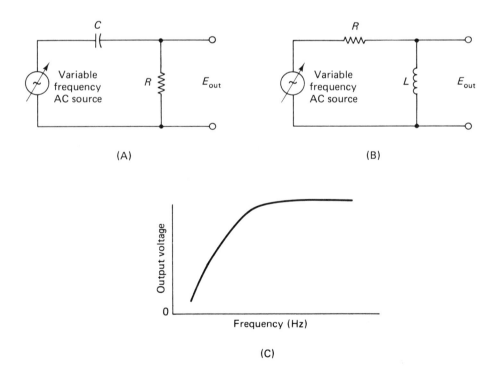

(A)

(B)

(C)

FIGURE 9–12   High-pass filters. (A) Series RC high-pass filter. (B) Series RL high-pass filter. (C) Frequency response curve of high-pass filter.

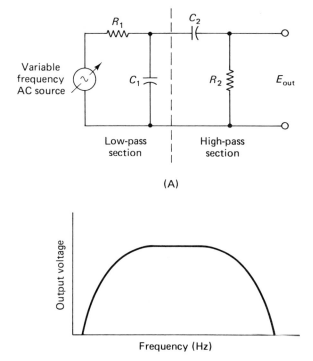

FIGURE 9–13 Bandpass filters. (A) Bandpass filter circuit configuration. (B) Frequency response curve of bandpass filter.

quencies and block low and high frequencies. R_1 and C_1 form a low-pass filter and R_2 and C_2 form a high-pass filter. The range of frequencies to be passed is determined by calculating the values of resistance and capacitance. Notice the shape of the frequency response curve of Figure 9–13B, showing that on the low-frequency end, as frequency goes down, voltage output goes down, and on the high-frequency end, as frequency increases, voltage output decreases.

Decibels

The *decibel* (dB) is commonly used to measure changes in power or voltage level. Decibel values are also used to plot frequency response curves, as shown in Figure 9–14. This unit of measure is determined by a logarithmic scale. Our ears, for example, respond to logarithmic changes in power level in terms of sound. A sound level that physically changes 10 times only appears as an increase of 1 to the human ear. Likewise, a sound level that increases 20 times appears as a change of 2. In this system a sound level or power level of 0 dB is considered to be at the threshold of hearing or reference level.

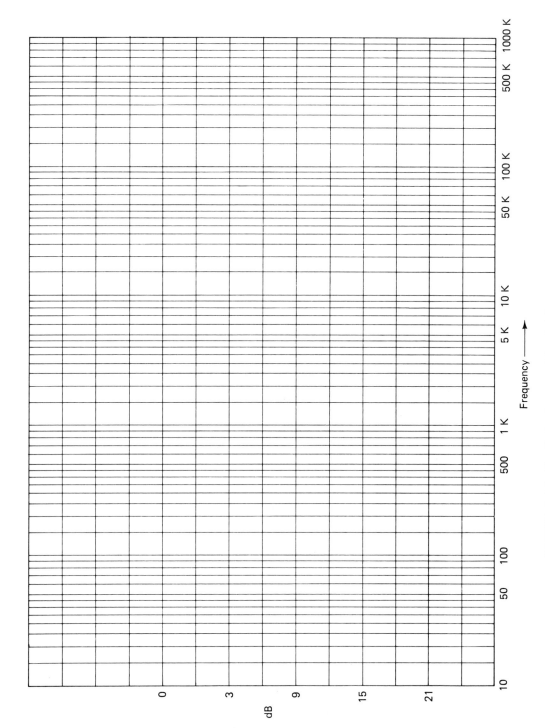

FIGURE 9–14 Decibel values used to plot frequency response.

The fundamental unit is the *bel*, which is derived from the surname of Alexander Graham Bell. The bel, however, is a rather large unit defined by the equation

$$\text{bel} = \log_{10} \frac{P_1}{P_2}$$

where P_1 and P_2 are comparisons of output power over input power and P_1 is greater than P_2.

A more practical expression of changes is represented by the decibel (dB). The term *deci* is the metric equivalent of 0.1 or $\frac{1}{10}$ of the fundamental unit. A dB is, therefore, $\frac{1}{10}$ of a bel, or 10 dB = 1 bel. The mathematical expression of a decibel is

$$\text{dB} = 10 \ \log_{10} \frac{P_1}{P_2} \qquad \text{where } P_1 > P_2$$

$$\text{dB} = 20 \ \log_{10} \frac{E_1}{E_2} \qquad \text{where } E_1 > E_2$$

The *logarithm* of a number is the power to which another number (called the *base*) must be raised to equal that number. A *common logarithm* is expressed in powers of 10, as illustrated by the following:

$$10^4 = 10,000 \quad \text{or log } 10,000 = 4$$
$$10^3 = 1000 \quad \text{or log } 1000 = 3$$
$$10^2 = 100 \quad \text{or log } 100 = 2$$
$$10^1 = 10 \quad \text{or log } 10 = 1$$
$$10^0 = 1 \quad \text{or log } 1 = 0$$

The logarithm of any number between 10 and 100 has a characteristic value of 1. The *characteristic* is an expression of the magnitude range of the number. Numbers between 100 and 1000 have a characteristic of 2, and numbers between 1000 and 10,000 have a characteristic of 3. Number values less than 1.0 are expressed as a negative characteristic. It is generally not customary in electronic applications to use negative characteristic values.

When a number is not an even multiple of 10, it must have a decimal part called the *mantissa*. The logarithm of a number such as 285 is expressed as 2.4548. The characteristic is 2 because 285 is between 100 and 1000. The mantissa of 285 is 0.4548. The mantissa is found from the Table of Common Logarithms in Appendix 1. To find this value, look down the vertical number column (N) to find the number 28. Move horizontally to the 5 column to find that the mantissa for 285 is 4548. The logarithm of 285 is therefore 2.4548. The characteristic is 2 and the mantissa is 0.4548. Logarithm values may more conveniently be determined on most calculators.

The mantissa is always the same regardless of the location of the decimal point. For example, the mantissa is the same for 2850, 285, 28.5, and 2.85. The only difference in these values is the characteristic. The mantissa for 285 is 0.4548. The logarithms of the four values listed are 3.4548, 2.4548, 1.4548, and 0.4548.

A sample problem using decibels follows. The gain of an amplifier which has an input of 25 mW and an output of 300 mW is determined as follows. The power gain is found by using the formula

$$dB = 10 \ log \ \frac{P_{out}}{P_{in}} = 10 \ log \ \frac{300 \ mW}{25 \ mW}$$
$$= 10 \ log \ 12$$
$$= 10 \times 1.0791$$
$$= 10.791$$

The voltage gain of an amplifier may also be expressed in decibel values. The decibel voltage gain formula is

$$dB = 20 \ log \ V_0/V_{in}$$

Notice that the logarithm of V_0/V_{in} is multiplied by 20 in this equation.

As an example of the decibel voltage gain equation, assume that an amplifier has an input voltage of $0.45V_{p\text{-}p}$ and the output voltage is $0.86V_{p\text{-}p}$. The voltage gain in decibels is

$$dB = 20 \ log \ \frac{V_0}{V_{in}}$$
$$= 20 \ log \ \frac{0.86V_{p-p}}{0.45V_{p-p}}$$
$$= 20 \ log \ 1.91$$
$$= 20 \times 0.2812$$
$$= 2.812$$

When the decibel value of an amplifier is known, the *power gain* or *voltage gain* may be determined by using inverse logarithms or *antilogarithms*. An *inverse logarithm* is the number from which a logarithm is derived. The process of finding an inverse logarithm is the reverse of finding a logarithm. The mantissa value can be found in Appendix 1. Then the number values which correspond to the mantissa value are determined. The decimal point is located by looking at the characteristic value. These values may also be easily determined by using the inverse logarithm function on a calculator.

As an example of using inverse logarithms, assume that an amplifier has a decibel value of $+3.5$. The power gain of the amplifier is found as follows:

$$dB = 10 \log \frac{P_1}{P_2}$$

$$3.5 = 10 \log \frac{P_1}{P_2}$$

$$0.35 = \log \frac{P_1}{P_2} \quad \textbf{(divide both sides of the equation by 10)}$$

$$P_1/P_2 = \text{inv log } 0.35 \quad \textbf{(find inverse logarithm)}$$

$$P_1/P_2 = 2.24$$

The value of 2.24 obtained in the example is the power ratio. An amplifier with a gain of $+3.5$ dB thus has a power gain of 2.24 to 1. The inverse logarithm value of the example is found by looking for a mantissa value of 3500 in the table. The corresponding digits are 224. The decimal point is placed between the 2s since the characteristic of 0 means that the number must be between 1 and 10.

Decibels are also used to express reduction in power or voltage levels. Reduction of input signal level in a circuit is called *attenuation*. A circuit which attenuates a signal is compared to an amplifier circuit in Figure 9–15. Notice that the decibel value is marked with a minus sign

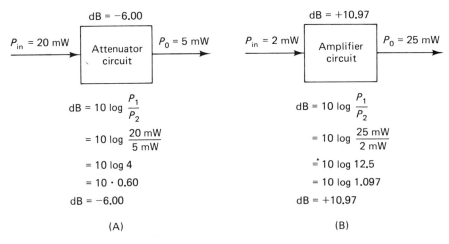

(A)   (B)

FIGURE 9–15   Comparison of (A) attenuator and (B) amplifier circuits.

when the circuit attenuates the input signal. A common example of attenuation occurs in coaxial cable or other signal transmission cable in which a reduction of signal occurs from input to output.

Using Decibels With Filter Circuits

Decibels are commonly used to plot frequency response curves for filter circuits. One example of a low-pass filter circuit is shown in Figure 9–16A. The procedure for plotting a frequency response curve for the low-pass circuit is shown in Figure 9–16B.

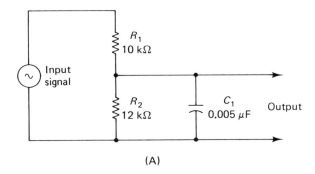

(A)

1. Find the Thevenin equivalent resistance (R_{Th}) of R_1 and R_2:

$$R_{Th} = R_1 \parallel R_2 = \frac{10\ K \times 12\ K}{10\ K + 12\ K} = 5.45\ k\Omega$$

2. Determine the 3-dB frequency using this formula:

$$f_{3\ dB} = \frac{1}{2\pi \cdot C \cdot R_{Th}} = \frac{1}{6.28 \cdot (0.005 \times 10^{-6}) \cdot (5.45 \times 10^3)}$$

$$= \frac{1}{1.7113 \times 10^{-4}} = 5843\ Hz$$

3. Find the 9-, 15-, and 21-dB frequencies:

$$f_{9\ dB} = 2 \cdot f_{3\ dB} = 2 \cdot 5843\ Hz = 11,686\ Hz$$
$$f_{15\ dB} = 4 \cdot f_{3\ dB} = 4 \cdot 5843\ Hz = 23,372\ Hz$$
$$f_{21\ dB} = 8 \cdot f_{3\ dB} = 8 \cdot 5843\ Hz = 46,744\ Hz$$

4. Label the points on a sheet of frequency response paper (see Figure 9-14).

5. Connect each of the points to form a low-pass frequency response curve.

(B)

FIGURE 9–16 Frequency response for a low-pass filter circuit.
(A) Circuit. (B) Procedure.

The selection of decibel values of 3, 9, 15, and 21 dB is standard for plotting frequency response in terms of voltage output of a circuit. These values are most easily interpreted by referring to Appendix 2. First, locate the 3-dB line. Notice that with a 3-dB reduction of a signal, the power output is approximately 0.5, or 50 percent, of the 0-dB reference level and the voltage is approximately 0.707, or 70.7 percent, of the 0-dB level. Since the power output of a circuit reduces to about 50 percent of its original value (0 dB), the 3-dB frequency is called the *half-power point*.

The values of the other decibel points plotted on a frequency response graph are as follows:

9 dB — power = 0.1259   Voltage = 0.3548

15 dB — power = 0.03162   Voltage = 0.1778

21 dB — power = 0.01581   Voltage = 0.0887

The selection of decibel points for a high-pass filter circuit is similar to the process used for low-pass filter circuits. A high-pass filter circuit and the procedure for plotting a frequency response curve are shown in Figure 9–17.

Bandpass filter circuits are a combination of low-pass and high-pass filter circuits. An example of a bandpass filter circuit and the procedure for plotting a frequency response curve are shown in Figure 9–18. Notice the use of the combined resistances $(R_1 + R_2)$ for the high-pass section and the use of the *Thevenin equivalent resistance* (R_{Th}) for the low-pass section. The 3-dB frequency on the low-frequency end of the response curve is called the *low cutoff frequency* (f_{LC}). The high-frequency 3-dB point is called the *high cutoff frequency* (f_{HC}).

RESONANT CIRCUITS

Resonant circuits are designed to pass a certain range of frequencies and block other frequencies. Such circuits have resistance, inductance, *and* capacitance. Figure 9–19 shows the two types of resonant circuits—*series resonant circuits* and *parallel resonant circuits*—and their frequency response curves, which show how circuit current and impedance vary with frequency.

Series Resonant Circuits

Series resonant circuits are series circuits which have inductance, capacitance, and resistance. Series resonant circuits offer a low impedance to some AC frequencies and a high impedance to other frequencies.

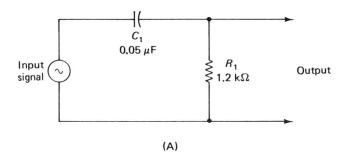

(A)

1. Determine the 3-dB frequency:

$$f_{3\ dB} = \frac{1}{2\pi \cdot R \cdot C} = \frac{1}{6.28 \times (1.2 \times 10^3) \times (0.05 \times 10^{-6})}$$

$$= \frac{1}{3.768 \times 10^{-4}} = 2654\ Hz$$

2. Find the 9-, 15-, and 21-dB frequencies:

$$f_{9\ dB} = f_{3\ dB} \div 2 = 2654 \div 2 = 1327\ Hz$$
$$f_{15\ dB} = f_{3\ dB} \div 4 = 2654 \div 4 = 663.5\ Hz$$
$$f_{21\ dB} = f_{3\ dB} \div 8 = 2654 \div 8 = 332\ Hz$$

3. Label each of the points on a sheet of frequency response paper.

4. Connect each of the points to form a high-pass frequency response curve.

(B)

FIGURE 9–17 Frequency response for a high-pass filter circuit. (A) Circuit. (B) Procedure.

They are used to select or reject frequencies of a frequency range applied
to the input of a circuit.
     The voltage across an inductor and a capacitor in an AC series cir-
cuit are in direct opposition to each other (180° out of phase) and cancel
each other out. The frequency applied to a series resonant circuit affects
inductive reactance (X_L) and capacitive reactance (X_C). At a specific input frequency, X_L of a series RLC circuit equals X_C. The voltage across
the inductor (E_L) and capacitance (E_C) in the circuit are then equal. The
total reactive voltage (E_X) is 0 V at this frequency. The reactance of the
inductor and the capacitor cancel each other at this frequency. The total
reactance (X_T) of the circuit ($X_L - X_C$) is zero. The impedance (Z) of the
circuit is then equal to the resistance (R).

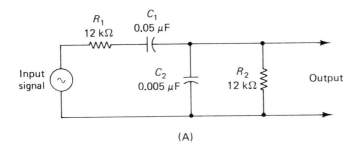

(A)

1. Determine the 3-, 9-, 15-, and 21-dB frequencies for the high-pass section:

$$f_{3\,dB} = \frac{1}{2\pi \cdot R \cdot C} = \frac{1}{6.28 \times (24 \times 10^3) \times (0.05 \times 10^{-6})}$$

$$= \frac{1}{7.536 \times 10^{-3}} = 133 \text{ Hz } (f_{LC})$$

$f_{9\,dB} = f_{3\,dB} \div 2 = 66.5 \text{ Hz}$

$f_{15\,dB} = f_{3\,dB} \div 4 = 33.25 \text{ Hz}$

$f_{21\,dB} = f_{3\,dB} \div 8 = 16.6 \text{ Hz}$

2. Determine the 3-, 9-, 15-, and 21-dB frequencies for the low-pass section:

$$f_{3\,dB} = \frac{1}{2\pi \cdot R_{Th} \cdot C} = \frac{1}{6.28 \times (6 \times 10^3) \times (0.005 \times 10^{-6})}$$

$$= \frac{1}{1.884 \times 10^{-4}} = 5308 \text{ Hz } (f_{HC})$$

$f_{9\,dB} = f_{3\,dB} \times 2 = 10,616 \text{ Hz}$

$f_{15\,dB} = f_{3\,dB} \times 4 = 21,232 \text{ Hz}$

$f_{21\,dB} = f_{3\,dB} \times 8 = 42,464 \text{ Hz}$

3. Label each of the points on a sheet of frequency response paper.

4. Connect each of the points to form a bandpass frequency response curve.

(B)

FIGURE 9–18 Frequency response for a bandpass filter circuit. (A) Circuit. (B) Procedure.

The frequency at which $X_L = X_C$ in a circuit is called *resonant frequency*. To determine the resonant frequency (f_r) of the circuit, use the formula $f_r = \dfrac{1}{2\pi \times \sqrt{L \times C}}$ In this formula L is in henries, C is in farads, and f_r is in hertz. Notice that as either inductance or capacitance increases, resonant frequency decreases. When the resonant frequency is applied

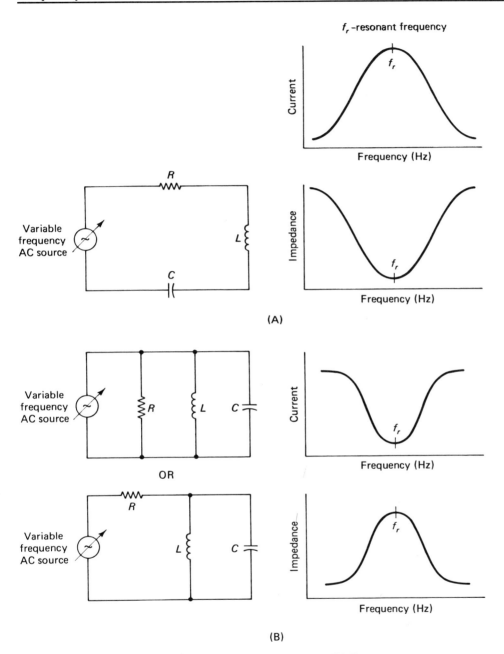

FIGURE 9–19   Resonant circuits. (A) Series resonant. (B) Parallel resonant.

to a circuit, a condition called *resonance* exists. Resonant frequency is calculated in the same way for series and parallel circuits.

Resonance for a series circuit causes the following conditions to exist: (1) $X_L = X_C$, (2) X_T is equal to zero, (3) $E_L = E_C$, (4) total reactive voltage (E_X) is equal to zero, (5) $Z = R$, (6) total current (I_T) is maximum, and (7) phase angle (θ) is 0°. These are the basic characteristics of all series resonant circuits. Remember that these characteristics apply to series *RLC* circuits.

The ratio of reactance $(X_L$ or $X_C)$ to resistance (R) at resonant frequency of a circuit is called *quality factor (Q)*. This ratio is used to determine the range of frequencies or *bandwidth (BW)* a resonant circuit will pass. A sample series resonant circuit problem is shown in Figure 9–20. The frequency range that a resonant circuit will pass (BW) is found by using steps 5 and 6.

The frequency cutoff points are at approximately 70 percent of the maximum output voltage. These are called the *low-frequency cutoff* (f_{LC}) and *high-frequency cutoff* (f_{HC}). The bandwidth of a resonant circuit is determined by the Q of the circuit. The circuit Q is determined by the ratio X_L and X_C to R (X_L/R). Resistance is the primary factor which determines bandwidth, as summarized by the following: (1) when R is increased, Q decreases, since $Q = X_L/R$; (2) when Q decreases, bandwidth increases, since $BW = f_r/Q$; and (3) when R is increased, bandwidth increases.

Two series resonant circuit response curves are shown in Figure 9–21. These curves show the effect of resistance on bandwidth. The curve of Figure 9–21B has *high selectivity*, meaning that a resonant circuit with this response curve would select a small range of frequencies. Selectivity is very important for radio and television tuning circuits and other electronic applications.

Parallel Resonant Circuits

Parallel resonant circuits have characteristics similar to series resonant circuits. The electrical properties of parallel resonant circuits are somewhat different, but they accomplish the same purpose as series circuits. This circuit is a parallel combination of L and C used to select or reject AC frequencies.

With the resonant frequency (f_r) applied to a parallel resonant circuit, the following conditions occur: (1) $X_L = X_C$; (2) $X_T = 0$; (3) $I_L = I_C$; (4) $I_X = 0$, so circuit current is minimum; (5) $Z = R$ and is maximum; and (6) phase angle $\theta = 0°$.

The calculations used for parallel resonant circuits are similar to those for series circuits. However, quality factor Q is found for parallel circuits by using the formula $Q = R/X_L$, and current values are used

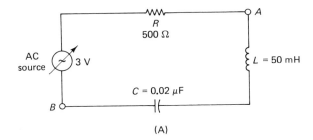

(A)

1. Find resonant frequency (f_r):

$$f_r = \frac{1}{2\pi \times \sqrt{L \times C}} = \frac{1}{6.28 \times \sqrt{(50 \times 10^{-3}) \times (0.02 \times 10^{-6})}}$$

$$= \frac{1}{6.28 \times \sqrt{1.0 \times 10^{-9}}} = \frac{1}{1.9 \times 10^{-4}} = 5035 \text{ Hz}$$

2. Find X_L and X_C at resonant frequency:

$X_L = 2\pi \cdot f \cdot L = 6.28 \times 5,035 \times 0.050 = 1581 \ \Omega$

(X_L easier to calculate.)

3. Find quality factor (Q):

$$Q = \frac{X_L}{R} = \frac{1581}{500} = 3.162$$

4. Find bandwidth (BW):

$$BW = \frac{f_r}{Q} = \frac{5035 \text{ Hz}}{3.162} = 1592 \text{ Hz}$$

5. Find low-frequency cutoff (f_{LC}):

$f_{LC} = f_r - \frac{1}{2} BW = 5035 - 796 = 4239 \text{ Hz}$

6. Find high-frequency cutoff (f_{HC}):

$f_{HC} = f_r + \frac{1}{2} BW = 5035 + 796 = 5831 \text{ Hz}$

7. Find I_T at resonant frequency:

$$I_T = \frac{E_A}{R} = \frac{3 \text{ V}}{500 \ \Omega} = 0.006 \text{ A} = 6 \text{ mA}$$

8. Find E_R at resonant frequency:

$E_R = I \times R = 0.006 \text{ A} \times 500 \ \Omega = 3 \text{ V}$

9. Find E_L and E_C at resonant frequency:

$E_L = E_C = I \cdot X_L = 0.006 \text{ A} \times 1581 \ \Omega = 9.49 \text{ V}$

Note that this voltage exceeds source voltage — this is called "voltage magnification"

(B)

FIGURE 9–20 Series-resonant circuit problem. (A) Circuit with component values. (B) Procedure to determine circuit values.

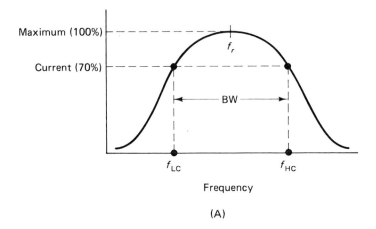

(A)

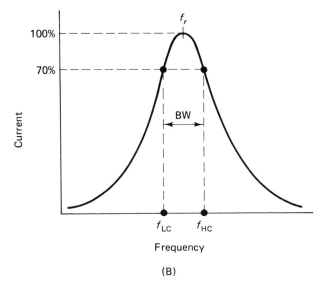

(B)

FIGURE 9–21   Effect of resistance on bandwidth and selectivity of resonant circuit. (A) High resistance (low selectivity). (B) Low resistance (high selectivity).

rather than reactive voltage drops. Figure 9–22 shows a parallel resonant circuit and an example calculation.

In this chapter oscillators and frequency-sensitive circuit applications have been discussed. Answer the following review questions, solve the problems, and perform the suggested laboratory activities to gain a better understanding of the contents of this chapter.

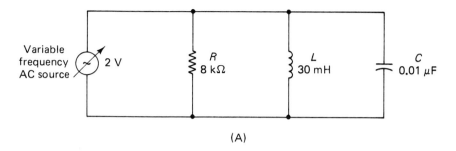

(A)

1. Find resonant frequency (f_r):

$$f_r = \frac{1}{2\pi \sqrt{L \times C}} = \frac{1}{6.28 \times \sqrt{(30 \times 10^{-3}) \times (0.01 \times 10^{-6})}}$$

$$= \frac{1}{1.0877 \times 10^{-4}} = 9194 \text{ Hz}$$

2. Find X_L and X_C at resonant frequency:

$$X_L = 2\pi \cdot f \cdot L = 6.28 \times 9194 \times (30 \times 10^{-3}) = 1732 \ \Omega$$

3. Find quality factor Q:

$$Q = \frac{R}{X_L} = \frac{8000 \ \Omega}{1732 \ \Omega} = 4.62$$

4. Find bandwidth (BW):

$$BW = \frac{f_r}{Q} = \frac{9194 \text{ Hz}}{4.62} = 1990 \text{ Hz}$$

5. Find low cutoff frequency (f_{LC}):

$$f_{LC} = f_r - \tfrac{1}{2} BW = 9194 - 995 = 8199 \text{ Hz}$$

6. Find high cutoff frequency (f_{HC}):

$$f_{HC} = f_r + \tfrac{1}{2} BW = 9194 + 995 = 10{,}189 \text{ Hz}$$

7. Find I_R at resonant frequency:

$$I_R = \frac{E_A}{R} = \frac{2 \text{ V}}{8 \text{ k}\Omega} = 0.25 \text{ mA}$$

8. Find I_L and I_C at resonant frequency:

$$I_L = \frac{E_A}{X_L} = \frac{2 \text{ V}}{1732 \ \Omega} = 1.15 \text{ mA}$$

(B)

FIGURE 9–22   (A) Parallel resonant circuit. (B) Problem-solving procedure.

REVIEW

1.   What is feedback?

2.   What are two types of feedback?

3.   How does feedback affect amplification gain?

4.   What is audio frequency?

5.   What is radio frequency?

6.   Why are the values of inductors and capacitors very critical in high-frequency amplifiers?

7.   What methods are used to couple one amplification circuit to another?

8.   What are the advantages of transformer coupling?

9.   What are oscillators?

10.   How are oscillators classified?

11.   What is a tank circuit?

12.   What is the flywheel effect?

13.   What determines the frequency of the waveform produced by the Hartley oscillator?

14.   Why doesn't the negative feedback associated with the Hartley oscillator prevent oscillation?

15.   What is an astable multivibrator?

16.   What turns the transistors of an astable multivibrator on and off?

17.   What controls the number of square waves per second that are produced by the astable multivibrator?

18.   What is a free-running oscillator?

19.   What is a bistable multivibrator?

20.   How does the bistable multivibrator differ from the astable multivibrator?

21.   How can a bistable multivibrator be triggered?

22.   Explain the purpose of each of the three basic types of filter circuits.

23.   What is a frequency response curve?

24.   What is a resonant circuit?

25. What is a decibel?

26. Explain the process of finding a logarithm of a number.

27. What is the difference between an amplifier and an attenuator?

28. Explain the process of determining a number by using an inverse logarithm.

29. What are the corresponding voltage output values of the following decibel levels: 3, 9, 15, and 21 dB?

30. How is value of the 3-dB frequency of a filter circuit found?

31. What is meant by low cutoff frequency?

32. What is meant by high cutoff frequency?

33. List some characteristics of series resonant circuits.

34. List some characteristics of parallel resonant circuits.

35. How is the bandwidth of a resonant circuit determined?

36. What is meant by the quality factor (Q) of a circuit?

PROBLEMS

1. A circuit has a loss of -16 dB. What power ratio corresponds to this loss?

2. The input to powerline is 220 mW. The power delivered at the end of the line is 40 mW. What is the attenuation in decibels?

3. A power input of an audio-frequency amplifier is 20 mW. The output is 120 mW. What is the amplification in decibels?

4. What is the ratio of output power to input power if there is a gain of 15 dB?

5. A coaxial transmission line has a power loss of -22 dB. Determine the power ratio.

6. Refer to Figure 9–16A. Use values of $R_1 = 20$ kΩ, $R_2 = 15$ kΩ, and $C_1 = 0.004$ μF for a low-pass filter circuit. Determine the 3-, 9-, 15-, and 21-dB frequencies.

7. Refer to Figure 9–14A. Use values of $R_1 = 2$ kΩ and $C_1 = 0.03$ pF for a high-pass filter circuit. Determine the 3-, 9-, 15-, and 21-dB frequencies.

8. Refer to Figure 9–18A. Use values of $R_1 = 10$ kΩ, $R_2 = 6$ kΩ, $C_1 = 0.04$

μF, and $C_2 = 0.0025$ μF for a bandpass filter circuit. Determine the low and high 3-, 9-, 15-, and 21-dB frequencies.

9. Refer to the series resonant circuit of Figure 9–20A. Use values of $R = 1$ kΩ, $L = 100$ mH, $E_{in} = 2$ V, and $C = 0.025$ μF. Determine the following:

a) resonant frequency

b) X_L at resonant frequency

c) quality factor

d) bandwidth

e) low-frequency cutoff

f) high-frequency cutoff

g) I_T at resonant frequency

h) E_R at f_r

i) E_L at f_r.

10. Refer to the parallel resonant circuit of Figure 9–22A. Use values of $R = 5$ kΩ, $L = 20$ mH, $C = 0.02$ μF, and $E_{in} = 1$ V. Determine the following:

a) f_r

b) X_L at f_r

c) Q

d) BW

e) f_{LC}

f) f_{HC}

g) I_R at f_r

h) I_L at f_r.

SUGGESTED LABORATORY ACTIVITIES

1. Construct the circuit illustrated in Figure 9–23 and draw the output waveforms as observed on the oscilloscope.

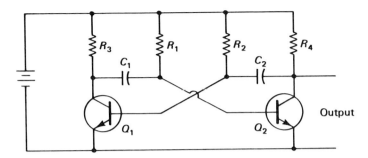

FIGURE 9–23   Multivibrator circuit.

2. Low-pass filter circuit.

    a) Contruct a low-pass filter circuit similar to the one shown in Figure 9–16A. Use values of R_1, R_2, C_1, and E_{in} as specified by the instructor. Use a variable-frequency audio oscillator to supply the input signal.

    b) Calculate the 3-dB frequency.

    c) Set the audio oscillator on a low frequency.

    d) With an oscilloscope adjust the vertical height of the output signal to a convenient reference level.

    e) Increase the frequency of the audio oscillator while observing the height of the oscilloscope signal. The 3-, 9-, 15-, and 21-dB values are found as follows:

$$3\ db = 70\% \quad \text{of original signal height}$$

$$9\ db = 35\% \quad \text{of original signal height}$$

$$15\ db = 17.5\% \ \text{of original signal height}$$

$$21\ db = \ 9\% \quad \text{of original signal height}$$

    f) Determine the frequencies which correspond to the decibel values in step e and plot a frequency response curve on semi-log paper such as shown in Figure 9–14.

3. High-pass filter circuit.

    a) Construct a high-pass filter circuit similar to the one shown in Figure 9–17A. Use values of R_1, C_1, and E_{in} as specified by the instructor. Use a variable-frequency audio oscillator as the input signal source.

b) Calculate the 3-dB frequency.

c) Set the audio oscillator on a high frequency.

d) Repeat steps d, e, and f of Lab 1. The frequency of the audio oscillator should be decreased to find the 3-, 9-, 15-, and 21-dB frequencies.

4. Bandpass filter circuit.

a) Construct a bandpass filter circuit similar to the one shown in Figure 9–18A. Use values of R_1, R_2, C_1, C_2, and E_{in} as specified by the instructor. Use a variable-frequency audio oscillator as the input signal source.

b) Calculate the low and high 3-dB frequencies.

c) Set the audio oscillator on an intermediate frequency.

d) Connect an oscilloscope across the output of the circuit.

e) Adjust the audio oscillator until maximum vertical deflection occurs. This should be the reference level (0 dB).

f) Increase, then decrease, the frequency of the audio oscillator to find the low and high 3-, 9-, 15-, and 21-dB frequencies, as explained in Lab 2.

g) Plot a frequency response curve for the bandpass filter circuit.

5. Series resonant circuit.

a) Construct a series resonant circuit similar to the one shown in Figure 9–20A. Use values of R, L, C, and E_{in} as specified by the instructor. Use a variable-frequency audio oscillator as the AC source.

b) Calculate (1) f_r, (2) X_L at f_r, (3) Q, (4) BW, (5) f_{LC}, (6) f_{HC}, (7) I_T at f_r, (8) E_R at f_r, and (9) E_1 at f_r.

c) Connect an oscilloscope across points A and B of the circuit.

d) Adjust the audio oscillator until the minimum vertical deflection occurs (the measured resonant frequency).

e) Use the proper procedure to plot a frequency response curve for the series resonant circuit.

6. Parallel resonant circuit.

a) Construct a parallel resonant circuit similar to the one shown in Figure 9–24. Use values of R, L, C, and E_{in} close to those of the figure.

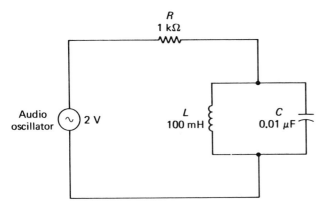

FIGURE 9–24 Parallel resonant circuit.

b) Calculate (1) f_r, (2) X_L at f_r, (3) Q, (4) BW, (5) f_{LC}, and (6) f_{HC}.

c) Connect an oscilloscope across the *LC* tank circuit. Adjust the audio oscillator until minimum vertical deflection occurs (the measured resonant frequency).

d) Plot a frequency response curve for the parallel resonant circuit.

CHAPTER 10
Optoelectronic Devices

Some optoelectronic devices are sensitive to light energy and react to light by changing resistance, emitting electrons, or developing voltage. Other optoelectronic devices actually emit light when placed under specified electrical conditions.

The applications of optoelectronic devices in the manufacturing industry are numerous. Optoelectronic devices, along with appropriate circuitry, are used to count, sense, scan, inspect, and in some instances totally control the manufacturing process.

In this chapter the characteristics of photoconductive, photovoltaic, and photoemissive devices will be examined. Likewise, the characteristics of solid-state lamps, as well as certain specialized electronic devices (including the light-activated silicon-controlled rectifier (LASCR), photodiode, and phototransistor) will be studied.

NATURE OF LIGHT

Light is a source of energy that travels at a number of electromagnetic frequencies as measured in *angstroms* (Å). An angstrom unit is indicative of the wavelength or frequency of light and is one-tenth of a *nanometer* (nm). Generally, light is divided into three distinct bands of frequencies that are classified as infrared, visible, and ultraviolet. Light classified as *visible* is visible to the human eye and usually ranges in frequencies from about 4000 to 7000 Å, as illustrated in Figure 10–1.

Light intensity is expressed in *lumens* and deals with the amount of light striking a surface of specified size, as provided by a standard

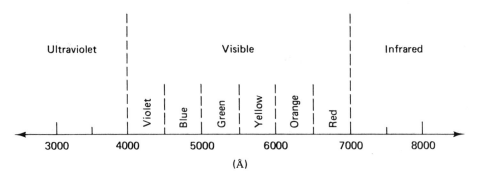

FIGURE 10–1   Frequencies of light energy.

light source from a specified distance. A standard 60-W incandescent lamp is rated as providing an average of 860 lm of light.

PHOTOCONDUCTIVE DEVICES

Photoconductive devices are sometimes called *photoresistive devices* because light brings about a change in their internal resistance. Generally, in the presence of light the resistance of the device decreases, while the absence of light causes the resistance of the device to increase.

Cadmium sulfide and cadmium selenide are two widely used materials in the manufacture of photoconductive devices, called *photoconductive (PC) cells*. These materials exhibit a unique characteristic, insomuch as light actually reduces their resistance. Light is a form of energy and causes the valence electrons of cadmium sulfide (CdS) to change energy levels, leave their orbital path, and become free electrons. An increase in the number of free electrons of any material will increase that material's conductivity, thus lowering the material's resistance. Figure 10–2 illustrates the typical construction, physical characteristics, and electrical symbol of a PC cell.

Figure 10–3 illustrates the electrical characteristics of a typical PC cell. The greater the light intensity up to a point, the less the PC cell's resistance. A PC cell's *"dark" resistance* represents its resistance in total darkness and may equal several hundred thousand ohms. With enough light intensity, the PC cell's resistance may be lowered to a few ohms.

PC cells, along with appropriate circuitry, are used to sense and measure light intensity. These cells find many industrial applications in the control of automated machinery. (PC cell applications are examined in Chapter 11.)

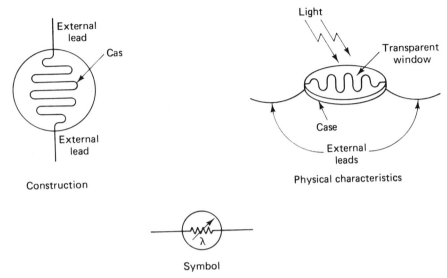

FIGURE 10–2   Photoconductive cell.

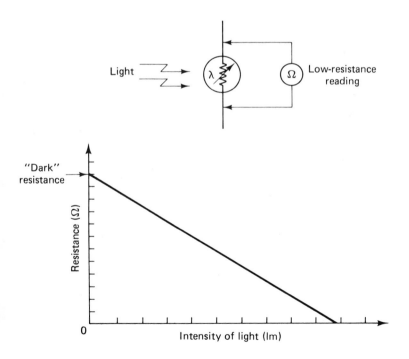

FIGURE 10–3   Characteristics of photoconductive cell.

Manufacturer's specifications associated with the photoconductive cell include maximum voltage, maximum current, maximum power dissipation, information concerning resistance versus light intensity, and response time. Usually these specifications are stated with the temperature of the device at 25 °C. The *maximum voltage specification* indicates the largest voltage that can be permitted to exist across the cell at any instant. The *maximum current specification* represents the largest amount of current that can pass through the device, while the *maximum power dissipation* indicates the PC cell's maximum combinations of voltage and current $(P = IE)$. Data concerning resistance versus light show the internal resistance caused by specific degrees of illumination. The response time is usually stated in microseconds and represents how quickly the change in resistance takes place.

PHOTOVOLTAIC DEVICES

Photovoltaic devices, more commonly called *photovoltaic (PV) cells*, or *solar cells*, are devices that cause a voltage to be generated when in the presence of light. Generally, as light intensity increases (up to a point) the voltage generated by the PV cell increases.

PV cells are generally constructed from selenium or silicon and are often called selenium or silicon cells. Selenium is more responsive to visible light while silicon responds better to infrared light. PV cells constructed of silicon are more efficient in converting light to voltage than are selenium cells. Figure 10–4 illustrates the physical characteristics and electrical symbol for the silicon PV cell.

The silicon PV cell exhibits the electrical characteristics of any PN junction diode when in the dark. That is, it conducts when forward biased and is nonconductive when reverse biased. When the PV cell is

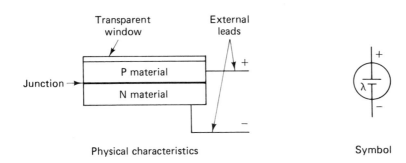

FIGURE 10–4   Photovoltaic cell.

exposed to light, this additional energy generates holes and electrons. The holes tend to move into the P material while the electrons move into the N material. This creates an excessive number electrons to exist in the N material and a deficiency of electrons in the P material. This creates a voltage potential between the P and N materials. The N material is negative with respect to the P material. An increase in light intensity causes the PV cell's voltage to increase. Figure 10–5 illustrates the electrical characteristics of a typical silicon PV cell. Notice that an increase in light causes the voltage produced by the PV cell to increase up to a point. This point is considered the PV cell's saturation point, where an increase in light intensity does not cause an increase in voltage. PV cells, along with appropriate circuitry, are used in many industrial applications (see Chapter 11).

Many manufacturers simply specify a PV cell's *output voltage* and *current* for a specific illumination level, along with the PV cell's *power rating*. These ratings are generally stated in millivolts, milliamperes, and milliwatts at 25 °C.

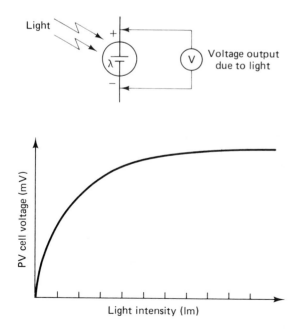

FIGURE 10–5 Characteristics of PV cell.

PHOTOEMISSIVE DEVICES

Photoemissive (PE) devices are those devices that actually "give off," or emit, electrons in the presence of light. An increased light intensity results in an increase in electrons emitted. Photoemissive devices are classified as either photovacuum or photogaseous tubes.

Chapter 1 discussed how light causes electron flow in a vacuum and a gas. In Chapter 2 electron flow in vacuum and gaseous diode tubes that is caused by heat was examined. Electron flow in both photovacuum and photogaseous tubes is caused and controlled by the same factors except heat is not present. Light intensity controls the number of electrons emitted by the light-sensitive cathode of both phototubes.

Figure 10–6 illustrates the electrical symbol for both the photovacuum and photogaseous tubes. The internal resistance of both tubes is very high; thus, the plate current caused by light in both tubes is usually described in microamperes.

PHOTOVACUUM TUBE

As with previously described vacuum tubes, the photovacuum tube's plate is positive and its cathode is negative. Light is caused to strike the tube's cathode, which is coated with a light-sensitive material. Electrons are emitted by the cathode and are attracted to the positive plate. As light intensity is increased, plate current is increased. Figure 10–7 shows a typical family of plate voltage curves for a photovacuum tube, along with a typical circuit.

The 25-MΩ load line of Figure 10–7 illustrates that as light intensity is increased, plate current increases and plate voltage decreases (points A, B, and C).

PHOTOGASEOUS TUBE

The photogaseous tube's operation depends upon light causing electrons to be emitted, resulting in the ionization of the gas within the tube. Because of the ionization of the gas within the photogaseous tube,

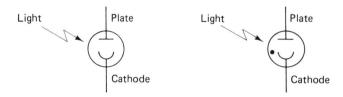

Photovacuum tube   Photogaseous tube

FIGURE 10–6   Photovacuum and photogaseous tubes.

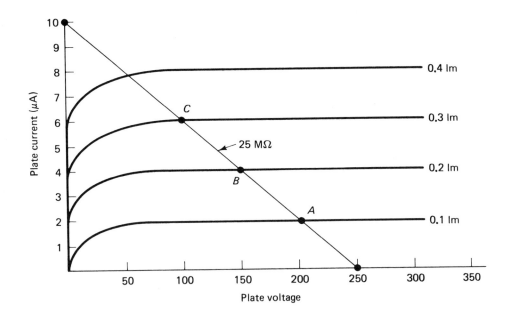

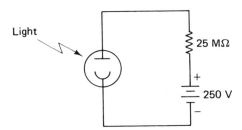

FIGURE 10–7   Family of plate voltage curves of photovacuum tube.

higher currents are associated with its conduction than are associated with the conduction of the photovacuum tube.

Both the photovacuum and photogaseous tube have generally been replaced by solid-state light sensors such as PC and PV cells.

PHOTOVACUUM MULTIPLIER TUBES

The multiplier tube is unique because it amplifies initial emission caused by light several hundreds of thousands of times. Figure 10–8 il-

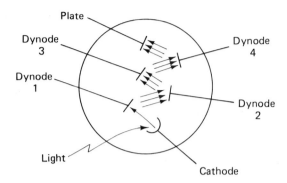

FIGURE 10–8   Photovacuum multiplier tube.

lustrates the internal arrangement of the tube's elements that cause this to happen.

Light strikes the photosensitive cathode and causes electrons to be emitted that are attracted to dynode 1. The electrons travel at such a velocity that when they reach dynode 1, secondary emission results. Electrons emitted by dynode 1 are greater in number than those emitted by the tube's cathode and are attracted to dynode 2, which is more positive than dynode 1. Electrons reaching dynode 2 cause secondary emission, increasing the electrons attracted by dynode 3, which is still even more positive than dynode 2.

Each additional dynode is more positive than its preceding one. The number of electrons emitted by each dynode increases progressively due to secondary emission. The number of electrons conducted by the plate is thousands of times greater than the number emitted by the cathode.

SOLID-STATE LAMPS (LEDs)

Solid-state lamps are devices constructed of semiconductive material that, when properly biased, gives off light. These are more commonly called *light-emitting diodes (LEDs)* and are used in groups to provide digital displays for many types of electronic systems. Figure 10–9 illustrates the construction characteristics and electrical symbol of a typical LED.

When current flows through any diode, recombination of holes and electrons takes place near the diode's junction. Electrons actually move because of the acquisition of additional energy from the forward-bias

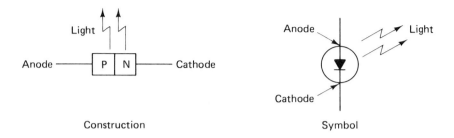

FIGURE 10–9   Construction characteristics of LED.

voltage source. When recombination takes place, this additional energy acquired by the electrons must be given up, usually in the form of heat. It was discovered that if special semiconductor material such as gallium arsenide or gallium phosphide is used to form PN junctions, the recombination of holes and electrons results in the emission of visible light. The color of the emitted light is controlled by the addition of selected phosphors.

The LED is extremely rugged and dependable. It is a low-powered device and lends itself very well to solid-state circuitry because of its rapid response time and low impedance. The LED is always forward biased to produce light. The graph illustrated in Figure 10–10 indicates the typical relation between light intensity and forward current flow.

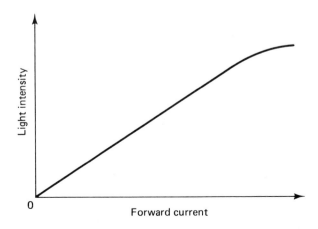

FIGURE 10–10   Current characteristics of LED.

SPECIAL PHOTOELECTRIC DEVICES

PHOTODIODE

The *photodiode* is a PN junction diode that exhibits the same forward conduction characteristics as most junction diodes. When the photodiode is reverse biased, light is used to control the reverse current. Figure 10–11 illustrates the electrical symbol and construction characteristics of a typical photodiode.

As light strikes the junction of the reverse-biased photodiode through the small "window" provided during the manufacturing process, reverse current flow is increased. An increase in light intensity results in an increase in reverse current. The light at the junction provides the additional energy to allow more carriers to break their bonds, allowing more recombination and more external current flow as measured in microamperes. A family of photodiode characteristic curves appears in Figure 10–12.

The *"dark" current* is the very small current that flows through the diode when reverse biased and in the absence of light. The dark current and the small amount of reverse current that flows normally when a junction is reverse biased are the same.

PHOTOTRANSISTOR

The *phototransistor* is a transistor that has its emitter-collector current controlled by light falling upon its base junction. The phototransistor may be of either the NPN or the PNP type. Figure 10–13 illustrates the construction and electrical symbols of the NPN phototransistor.

Figure 10–13 illustrates two symbols for the phototransistor. One symbol represents this device as having a base lead while the other does not. The base lead of the phototransistor is really not needed and is usually not provided by the manufacturer. When a base lead is provided, it is used for stabilization purposes.

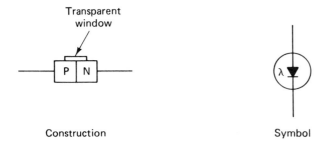

FIGURE 10–11   Photodiode construction characteristics.

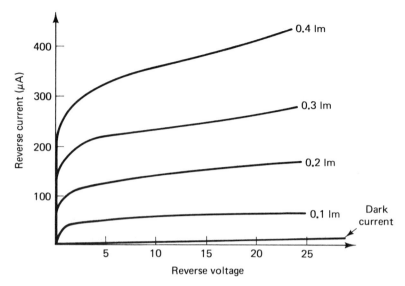

FIGURE 10–12   Photodiode characteristic curves.

When no light is applied to the phototransistor's base-emitter junction, very little base current flows, resulting in a very small emitter-collector current. With light applied to the transistor's base-emitter junction, electrons gain sufficient energy to break their bonds. This causes an increase in base-emitter current and results in an increase in emitter-collector current. As light intensity increases, base current increases,

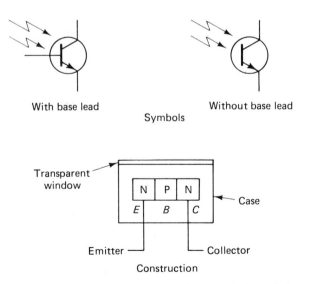

FIGURE 10–13   Construction characteristics of junction-type phototransistor.

causing collector current to increase. Figure 10–14 illustrates a family of collector curves and a representative circuit.

LIGHT ACTIVATED SILICON CONTROLLED RECTIFIER (LASCR)

The *LASCR* is an SCR that can be triggered by light in addition to conventional methods. Figure 10–15 illustrates the construction characteristics and the electrical symbol for the LASCR.

When light or a combination of light and gate current is used to trigger the LASCR, the light must be applied through the window to the gate-cathode junction. This allows more electrons to be freed for the reasons previously explained. If the light is intense enough for any given anode-cathode voltage, the LASCR is triggered on and will remain on, assuming a sufficient amount of holding current exists. Thus, the LASCR may be triggered on by gate current only, a combination of gate current and light, or by light only. The main advantage of the LASCR when used as a photoelectric device is that a very brief pulse of light activates the device and it remains activated as long as its anode remains positive, its cathode remains negative, and its holding current is sufficiently large.

In this chapter PC, PV, and PE devices, as well as certain specialized devices including photomultipliers, photodiodes, phototransistors, and LASCRs, have been studied. Additionally, in this chapter the LED, sometimes called a solid-state lamp, was introduced. Answer the following review questions and complete the suggested laboratory activities to gain a better understanding of the material in this chapter.

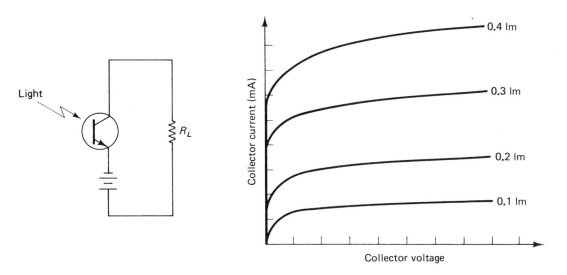

FIGURE 10–14   Family of collector curves.

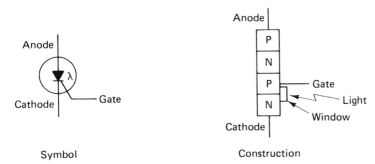

FIGURE 10–15   Construction characteristics of LASCR.

REVIEW

1.   What unit of measurement is used to express the frequency of light energy?

2.   What unit of measurement is used to indicate the intensity of light energy?

3.   What are the three distinct bands of light frequencies?

4.   What is the range of frequency for visible light?

5.   What are three distinct types of photoelectric devices?

6.   How does the PC cell differ from the PV cell?

7.   How does the PE device differ from the PV cell?

8.   How does the PE device differ from the PC cell? What do they have in common?

9.   What are the symbols for the PC, PV, and PE devices?

10.   How does the photovacuum and photogaseous tube differ? How are they similar?

11.   What two types of materials are used in the construction of the PC cell?

12.   What does the term *dark resistance* mean?

13.   What manufacturer's specifications are associated with the PC cell?

14.   What two materials are used to construct PV cells? Which of the materials is more efficient?

15.   What is the relation between light intensity and voltage of a PV cell?

16.   What manufacturer's specifications are associated with the PV cell?

17.   What is the relation between plate current and light intensity when discussing the photovacuum tube?

18.   What has replaced the phototubes?

19.   How does the photomultiplier amplify current?

20.   What is secondary emission?

21.   How does an LED cause light?

22.   What is the symbol for the LED, photodiode, phototransistor, and LASCR?

23.   What is the use of the LED?

24.   What is the relation between the forward current and the LED's light intensity?

25.   How is the photodiode's reverse current controlled by light?

26.   What is meant by the term *dark current*?

27.   How is the emitter-collector current of a phototransistor controlled by light?

28.   How many external leads are required to properly connect a phototransistor into a circuit?

29.   What is a LASCR?

30.   What are three methods that can be employed to trigger the LASCR?

SUGGESTED LABORATORY ACTIVITIES

1.   Construct the circuit illustrated in Figure 10–16 and record the PC's resistance for three different light intensities.

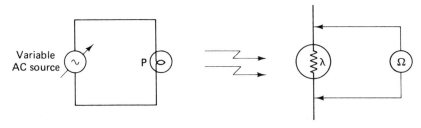

FIGURE 10–16   PC circuit.

2. Construct the circuit illustrated in Figure 10–17 and record the PV's voltage for three different light intensities.

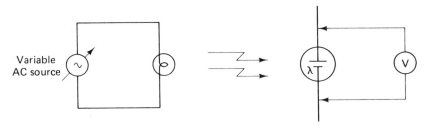

FIGURE 10–17 PV circuit.

3. Construct the circuit illustrated in Figure 10–18 and record the plate current for three different light intensities.

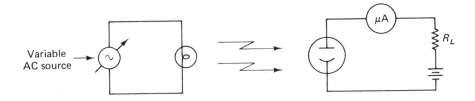

FIGURE 10–18 Photovacuum tube circuit.

4. Construct the circuit illustrated in Figure 10–19 and observe the light intensity of the LED for three different forward currents.

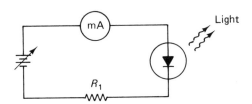

FIGURE 10–19 LED circuit.

5. Construct the circuit illustrated in Figure 10–20 and record the diode's reverse current for three different light intensities.

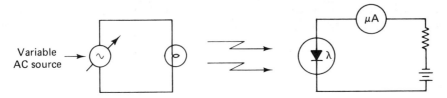

FIGURE 10–20   Photodiode circuit.

6.   Construct the circuit illustrated in Figure 10–21 and record the transistor's emitter-collector current for three different light intensities.

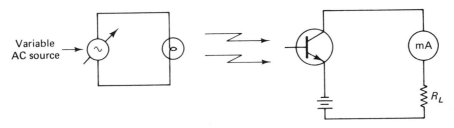

FIGURE 10–21   Phototransistor circuit.

7.   Construct the circuits illustrated in Figure 10–22 and show three different methods used to fire the LASCR.

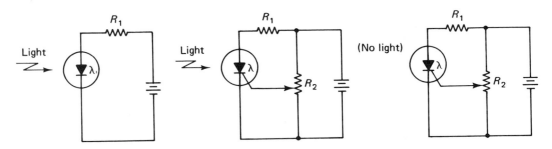

FIGURE 10–22   LASCR circuits.

CHAPTER 11
Optoelectronic
Circuits

Optoelectronic circuits are photosensitive devices, along with the appropriate components, connected in such a way as to scan, sense, count, measure, and otherwise control through the presence or absence of light energy.

In this chapter many types of optoelectronic circuits along with their applications will be examined. The important roles played by optoelectronic circuits in process control and instrumentation systems will be studied. Finally, optoelectronic circuits that are optically coupled to, but electrically isolated from, control circuits that are designed to perform specific functions will be discussed.

SCANNING, SENSING, AND PROCESS CONTROL CIRCUITS

To scan means to examine or check very intensively. A scanning circuit, through the use of light and optoelectronic circuitry, very carefully examines what it is supposed to examine, be it product, price, or recorded data. Figure 11–1 illustrates a scanning circuit that is used in the printing industry to scan and detect breaks in paper as it comes from the roll and is fed into the printing presses.

When the *photoconductive (PC) cell* of Figure 11–1 is in darkness (no break in the paper as it comes off the roll), its resistance is very high. This limits the current flow through R_1 and establishes a voltage drop

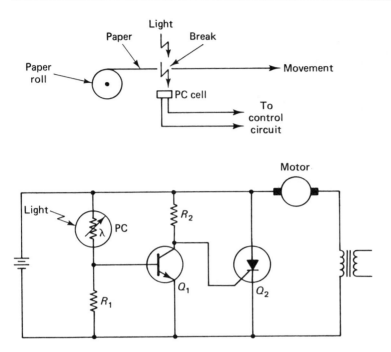

FIGURE 11–1   Photoelectric scanning circuit.

across R_1 that biases Q_1. The bias of Q_1 allows a very small emitter-collector current. This causes the emitter-collector voltage to be very high. Since the transistor's collector is positive as compared to its emitter, this places a high positive voltage on the gate of Q_2 (SCR). This causes Q_2 to be triggered on each time its anode is positive, thus causing the motor that is driving the mechanism that removes the paper from the roll to continue to run.

If the paper breaks, light strikes the window of the PC cell. This causes the cell's resistance to drop to a very low level. A reduction in the resistance of the PC cell allows more current to flow through R_1 and increases the voltage drop across R_1. This causes the base of Q_1 to become more positive, which results in a large increase in base current. The increase in base current causes the emitter-collector current to rise sharply. This increases the voltage drop across R_2 and causes the emitter-collector voltage to drop to a very low level. This causes Q_1's collector voltage to become less positive. When Q_1's collector voltage becomes less positive, Q_2 will no longer be triggered on and the motor controlling paper flow no longer operates. The system is restored when the break in the paper is mended.

Figure 11–2 illustrates a *photovoltaic (PV) cell,* with appropriate circuitry, that controls a conveyor and the filling of a container when light is interrupted by the container. This type of circuitry is widely used in industries where containers are automatically filled to contain a specific volume of material such as cereal, soap powder, dry pet food, or fertilizer.

When light strikes the PV cell of Figure 11–2 (no container present), it produces a voltage which makes the conductive channel of Q_1 (JFET) more narrow. This reduces source-drain current, which results in an increase in source-drain voltage. Since the drain of the JFET (Q_1) is positive with respect to the source, this causes Q_2 (SCR) to be triggered on, allowing the motor (1) that controls the conveyor to operate. This

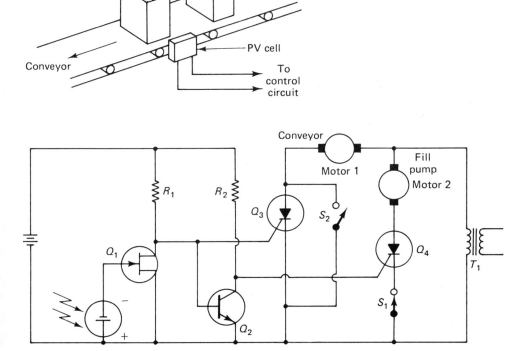

FIGURE 11–2   PV cell control circuit.

same source-drain voltage causes Q_2 (NPN transistor) to conduct heavily, reducing its collector voltage, which is positive, to a very low level. This serves to cause Q_4 (SCR) to remain off and the motor (2) that controls the filling operation likewise to remain off.

When a container moved by the conveyor blocks the light to the PV cell, the PV's voltage drops to zero and Q_1 conducts more heavily. This causes the drain voltage to be reduced to a very low level. This low level of source-drain voltage is not sufficient to trigger Q_3; thus, Q_3 (SCR) is turned off and motor 1 (conveyor control) does not operate, causing the conveyor to stop. This positions the container to be filled. Since the source-drain voltage of Q_1 is reduced to a very low level, this causes Q_2 (NPN transistor) to conduct less. This results in the collector-emitter voltage rising to a higher level. Since the collector of Q_2 is positive as compared to the emitter, Q_4 is triggered on, which causes the motor (2) that controls the "fill" operation to be operative.

S_1 and S_2 are pressure switches located under the conveyor. These switches react to pressure in an opposite fashion. When the container becomes filled, it weighs more and applies more pressure to S_1 and S_2. When the appropriate pressure is applied to these switches, due to an increased weight in the container because of the filling procedure, S_1 will open and S_2 will close. This causes Q_4 to be nonconductive and motor 2 (motor that controls the fill procedure) to be turned off. Likewise, S_2 is closed, causing motor 1 (motor that activates the conveyor) to be turned on. This stops the filling procedure and moves the container down the conveyor and restores light to the PV cell. When the pressure provided by the container is no longer applied to S_1 and S_2, S_1 closes and S_2 opens. The entire control circuit is recycled for the next container to be filled.

Figure 11–3 illustrates a *light-activated SCR (LASCR)*, along with appropriate circuitry, being used to scan flue tiles for an appropriate white lining material. Absence of the white lining results in the LASCR control circuit activating a mechanism that rejects those flue tiles that have no white linings. When light is reflected from the white lining, Q_1 (LASCR) is triggered on. This causes its positive anode voltage to drop to a low level, reducing the gate voltage of Q_2 (SCR). This causes Q_2 to remain off, allowing the motor that controls the rejecting mechanism to remain inoperative.

When no white lining is present, no light is reflected to the LASCR (Q_1); thus, it is nonconductive. This causes the anode voltage of Q_1 to rise to a much higher level, making the gate of Q_2 more positive. This triggers Q_2 on, resulting in the motor that controls the rejecting mechanism to be operative. The part without the white lining is rejected.

Figure 11–4 illustrates an optoelectronic circuit designed to keep a vat filled with liquid. As long as the vat remains unfilled, light causes the base-emitter junction of Q_1 to be forward biased. This causes an in-

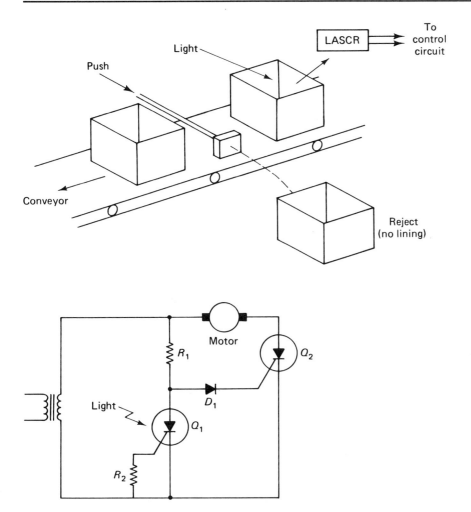

FIGURE 11–3   LASCR control circuit.

crease in Q_1's emitter-collector current, an increase in the voltage drop across R_3, and a decrease in emitter-collector voltage. This decrease in Q_1's collector voltage causes the base-emitter junction of Q_2 to become less forward biased, resulting in less emitter-collector current and an increase in emitter-collector voltage. This makes the gate of Q_3 more positive and Q_3 is triggered on. This causes the pump motor to be turned on, resulting in a continuous flow of liquid into the vat until it is filled.

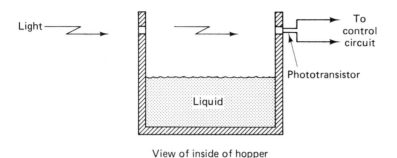

View of inside of hopper

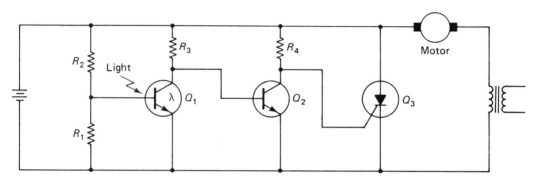

FIGURE 11–4   Phototransistor control circuit.

When the liquid level is sufficient to block light from Q_1, its conduction is decreased and its collector voltage rises. This causes Q_2 to conduct more heavily, resulting in a decrease in Q_2's collector voltage. This reduces Q_3's gate potential, causing it to be turned off. The pump motor likewise is turned off.

Figure 11–5 illustrates an LASCR used in a time delay circuit. Light striking Q_1 (LASCR) causes it to be conductive and provides a low resistive path across C_1. This prevents C_1 from charging to a voltage that is significant and prevents Q_2 from being triggered on.

When the product moves down the conveyor and blocks the light from Q_2, Q_1 becomes nonconductive and C_1 charges through D_2 and R_4. When the voltage across C_1 equals the zener voltage of D_1, D_1 becomes conductive. This applies a positive voltage to the gate of Q_2, Q_2 is triggered on, the solenoid valve is energized to open, and paint for color coding purposes is applied to the product. The amount of time delay between the light blockage and the application of the color code paint is controlled by the RC time constant of R_4 and C_1. The physical length of

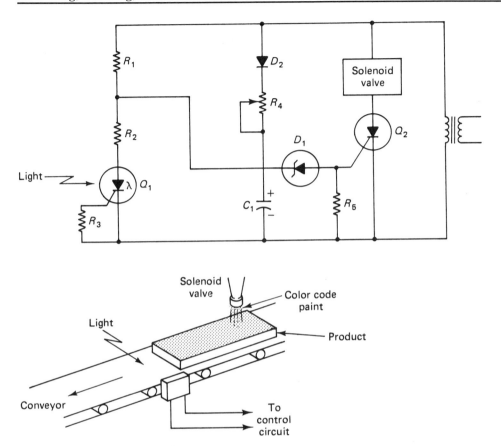

FIGURE 11–5   LASCR time delay circuit.

the product and the proper time to apply the paint should be considered when choosing values for C_1 and R_4.

When the product moves past the light source, light is restored to the LASCR (Q_1), C_1 is discharged, Q_2 becomes nonconductive, and the solenoid valve is closed. The entire cycle is repeated each time a product passes between the light and the LASCR.

Figure 11–6 illustrates an optoelectronic circuit that measures bulk material level in a hopper. When the hopper is empty, light strikes all PC cells and causes their collective resistances to be very low. This causes a maximum current flow through R_1, resulting in a voltage drop across R_1 that forward biases Q_1 and drives the transistor to saturation. This causes the collector voltage of Q_1 to decrease to near zero. Since the collector voltage of Q_1 controls the base voltage of Q_2, Q_2 is turned off.

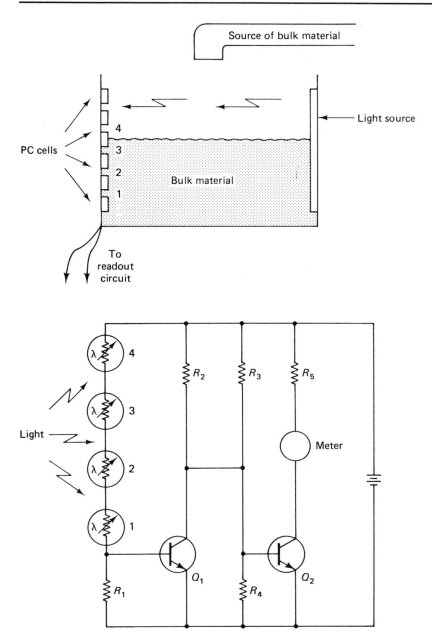

FIGURE 11–6   Bulk level sensor circuit.

There is very little, if any, collector-emitter current, and the meter indicates a reading of near zero.

As the hopper becomes filled, the PC cells are blocked from light and their collective resistances begin to increase. This reduces the current through R_1 as well as the voltage across R_1. Q_1 becomes less forward biased, its collector-emitter current decreases, and its collector voltage rises. This forward biases the base of Q_2, allowing collector-emitter current flow through the meter. The meter is appropriately calibrated to indicate relative bulk material level in the hopper.

OPTICAL COUPLING

When ambient light, such as sunlight, becomes a problem with the operation of optoelectronic circuits, a special modulated light system is employed that will respond only to pulsed light of a specific frequency. Figure 11–7 illustrates such a system. The LEDs are connected to the

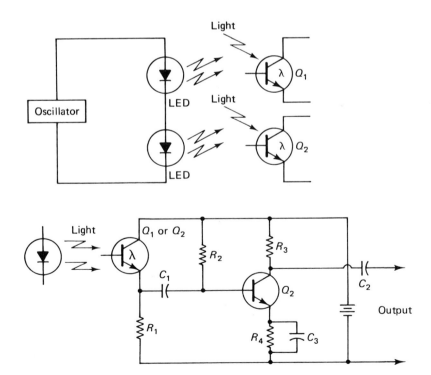

FIGURE 11–7   Optical coupling circuits.

output of an oscillator and provide pulsating light at the same frequency as the oscillator's output. Q_1 responds to this light frequency. The signal variation resulting across R_1 is coupled to the base of Q_2 through C_1. This results in the emitter-collector current of Q_2 exhibiting the same variations as the light produced by the LED. These variations appear as the output of Q_2 and are coupled through C_2 to additional amplification stages or control circuitry. LED pulse modulations can be separated from their detector-amplifier circuits by up to 200 feet and still be effective and reliable.

In this chapter applications of optoelectronic circuitry have been examined. Answer the following questions and perform the suggested laboratory activities as an aid in understanding optoelectronic circuits and their applications.

REVIEW

1. What are optoelectronic circuits and what are some of their advantages?

2. How could the circuit illustrated in Figure 11–1 be altered to cause the motor to operate when the PC cell is exposed to light rather than vice versa?

3. Why are S_1 and S_2 necessary in the circuit illustrated in Figure 11–2?

4. Could the PV cell illustrated in Figure 11–2 be replaced with a PC cell? Explain why or why not.

5. What is the purpose of Q_2 in Figure 11–2?

6. What causes Q_2 in Figure 11–3 to be turned off once it has been triggered on?

7. What would be the effect upon the operation of the circuit illustrated in Figure 11–5 if the value of R_4 or C_1 is altered?

8. Why is the length of the product illustrated in Figure 11–5 critical to the length of time delay?

9. Explain how the circuit illustrated in Figure 11–6 can measure the level of bulk material in the hopper.

10. What is a modulated light system and what are its advantages?

SUGGESTED LABORATORY ACTIVITIES

1. Construct the circuit illustrated in Figure 11–8. Measure and record the collector-emitter voltage and voltage drop across R_1 when the PC cell is in light and darkness.

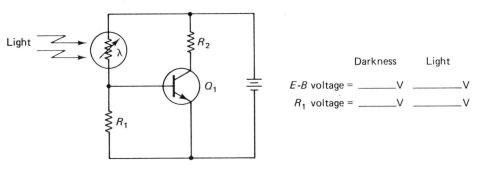

	Darkness	Light
E-B voltage =	_____V	_____V
R_1 voltage =	_____V	_____V

FIGURE 11–8 PC circuit.

2. Construct the circuit illustrated in Figure 11–9 and measure and record the source-drain voltage across the JFET when the PV cell is in light and darkness.

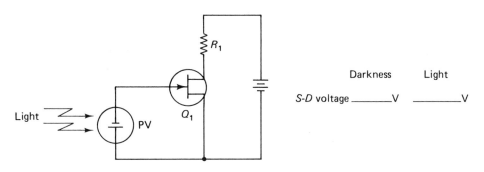

	Darkness	Light
S-D voltage	_____V	_____V

FIGURE 11–9 PV circuit.

3. Construct the circuit illustrated in Figure 11–10 and measure and record the voltage across C_1 when the LASCR is in light and darkness.

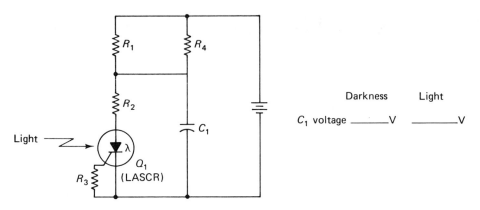

FIGURE 11–10   LASCR circuit.

4.   Construct the circuit illustrated in Figure 11–11 and measure and record the collector-emitter voltage of Q_1 with all PC cells in light, two PC cells in light, one PC cell in light, and all PC cells in darkness.

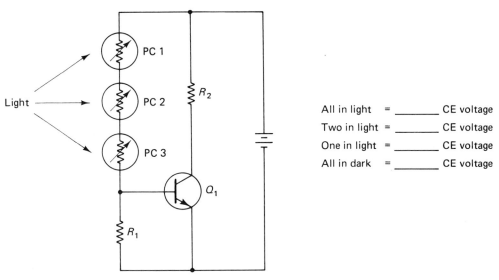

FIGURE 11–11   Bulk level circuit.

CHAPTER 12
Introduction to Digital Circuits

A circuit that employs a numerical signal in its operation is classified as a *digital circuit*. Computers, pocket calculators, digital instruments, and numerical control (NC) equipment are common applications of digital circuits. Practically unlimited quantities of digital information can be processed in short periods of time electronically. With operational speed of prime importance in electronics today, digital circuits are used more frequently.

In this chapter digital circuits are discussed. There are many types of digital circuits which have applications in electronics, including logic circuits, flip-flop circuits, counting circuits, and many others. The first sections of this chapter discuss the number systems which are basic to digital circuit understanding. The remainder of the chapter introduces some of the types of digital circuits.

DIGITAL NUMBER SYSTEM

The most common number system used today is the *decimal* system. In this number system ten digits are used for counting: 0, 1, 2, 3, 4, 5, 6, 7, 8, and 9. The number of digits of the system is called its *base*, or *radix*. The decimal system, therefore, has a base of 10.

Numbering systems have a place value, which refers to the placement of a digit with respect to others in the counting process. The largest digit that can be used in a specific place or location is determined by the base of the system. In the decimal system the first position to the left of the decimal point is called the *units* place. Any digit from 0 to 9 can be used in this place. When number values greater than 9 are used, they must be expressed in two or more places. The next position to the left of the units place is the 10s place in a decimal system. The number 99 is the largest digital value that can be expressed by two places in the decimal system. Each place added to the left extends the number value by a power of 10.

A number value of any base can be expressed by addition of weighted place values. The decimal number 2583, for example, is expressed as: $(2 \times 1000) + (5 \times 100) + (8 \times 10) + (3 \times 1)$. These values increase for each place extending to the left of the starting position or decimal point. These place or position factors can be expressed as powers of the base number. In the decimal system this is 10^3, 10^2, 10^1, and 10^0, with each succeeding place being expressed as the next larger power of base 10. Mathematically, each place value is expressed as the digit number times a power of the numbering system base. The decimal number 5362 is expressed in Figure 12–1.

The decimal number system is commonly used in our daily lives. Electronically, however, it is rather difficult to use. Each digit of a base 10 system would require a specific value associated with it. Electronically a system using this numbering method would not be practical.

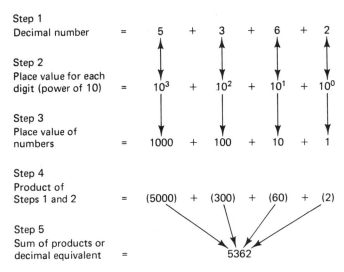

FIGURE 12–1   Expressing a decimal number.

BINARY NUMBER SYSTEM

Electronic digital systems are ordinarily of the *binary* type, which has 2 as its base or radix. The largest digital value that can be expressed by a specific place by this system is the number 1. Only the numbers 0 or 1 are used in the binary system. Electronically, the value of 0 can be expressed as a low-voltage value or no voltage. The number 1 can then be indicated by a voltage value larger than 0. Binary systems that use these voltage values are said to have *positive logic*. *Negative logic*, by comparison, has a voltage assigned to the 0 and no voltage value assigned to the number 1. Positive logic is used in this chapter.

The two operational states of a binary system, 1 and 0, are natural circuit conditions. When a circuit is turned off or has no voltage applied, it is in the off, or "0," state. An electrical circuit that has voltage applied is on, or in the "1" state. By using transistors or ICs it is electronically possible to change states in less than a microsecond. Electronic devices make it possible to manipulate millions of 0s and 1s in a second, and thus to process information quickly.

The basic principles of numbering used in decimal (base 10) numbers apply in general to binary numbers. The base of the binary system is 2, meaning that only the digits 0 and 1 are used to express place value. The first place to the left of the starting point, called the binary point, represents the units, or 1s, location. Places that follow to the left of the binary point are the powers of 2. Some of the digital values of numbers to the left of the binary point are $2^0 = 1$, $2^1 = 2$, $2^2 = 4$, $2^3 = 8$, $2^4 = 16$, $2^5 = 32$, $2^6 = 64$, etc.

When different number systems are used, they should have a subscript number to identify the base of the number system used. The number 100_2 is an example. This would be described as one-zero-zero instead of the decimal equivalent of one hundred.

The number 100_2 is equivalent to 4 in the base 10, or 4_{10}. Starting at the first digit to the left of the binary point, this number would have place value of $(0 \times 2^0) + (0 \times 2^1) + (1 \times 2^2)$ or $0 + 0_0 + 4_{10} = 4_{10}$. The conversion of a binary number to an equivalent decimal number is shown in Figure 12–2. In this method of conversion write down the binary number first. Starting at the binary point, indicate the decimal equivalent for each binary place location where a 1 is indicated. For each 0 in the binary number leave a blank space or indicate a 0. Add the place values and then record the decimal equivalent.

The conversion of a decimal number to a binary equivalent is achieved by repetitive steps of division by the number 2. When the quotient is even with no remainder, a 0 is recorded. When the quotient has a remainder, a 1 is recorded. The steps to convert a decimal number to binary number are shown in Figure 12–3.

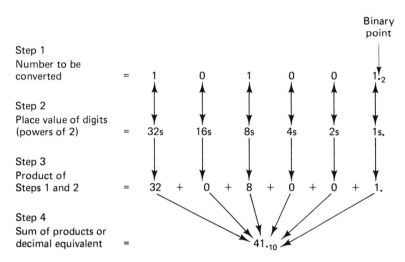

FIGURE 12–2   Binary-to-decimal conversion.

This conversion is achieved by writing down the decimal number (45_{10}). Divide this number by the base of the system (2). Record the quotient and remainder as indicated. Move the quotient of Step 1 to Step 2 and repeat the process. The division process continues until the quotient becomes zero. The binary equivalent is the remainder values in their last-to-first placement order.

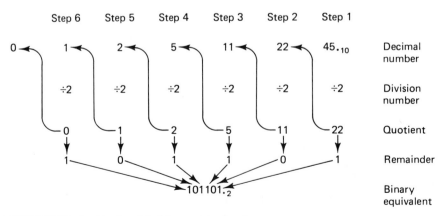

FIGURE 12–3   Decimal-to-binary conversion.

BINARY-CODED DECIMAL NUMBER SYSTEM

When large numbers are indicated by binary numbers, they are difficult to use. For this reason, the *binary-coded decimal (BCD) method* of counting was devised. In this system four binary digits are used to represent each decimal digit. To illustrate this procedure, the number 105_{10} is converted to a BCD number. In binary numbers $105_{10} = 1000101_2$.

To apply the BCD conversion process, the base 10 number is first divided into digits according to place values. See Figure 12–4. The number 105_{10} equals the digits 1-0-5. Converting each digit to binary would permit us to display this number as $0001\text{-}0000\text{-}0101_{BCD}$. Decimal numbers up to 999_{10} may be displayed by this process with only 12 binary numbers. The line between each group of digits is important when displaying BCD numbers.

The largest digit to be displayed by any group of BCD numbers is 9. Six digits of a number-coding group are not used at all in this system. Because of this, the octal (base 8) and the hexadecimal (base 16) systems were devised. Digital circuits process numbers in binary form but usually display them in BCD, octal, or hexadecimal form.

OCTAL NUMBER SYSTEM

The octal (base 8) number system is used to process large numbers by digital circuits. The *octal system* of numbers uses the same basic principles as the decimal and binary systems.

Given decimal number		$105._{10}$	
Step 1 Grouping of digits	(1)	(0)	(5)
Step 2 Conversion of each digit to binary group	(0001)	(0000)	(0101)
Step 3 Combine group values	$0001/0000/0101._{BCD}$		

FIGURE 12–4 Decimal-to-BCD conversion.

The octal number system has a radix, or base, of 8. The largest number displayed by the system before it changes the place value is 7. The digits 0, 1, 2, 3, 4, 5, 6, and 7 are used in the place positions. The place values of digits starting at the left of the octal point are the powers of eight: $8^0 =$ units, or 1s, $8^1 = 8$s, $8^2 = 64$s, $8^3 = 512$s, $8^4 = 4096$s, etc.

The process of converting an octal number to a decimal number is the same as that used in the binary-to-decimal conversion process. In this method, however, the powers of 8 are used instead of the powers of 2. Suppose that the number 382_8 is to be changed to an equivalent decimal number. The procedure is outlined in Figure 12–5.

Converting an octal number to an equivalent binary number is similar to the BCD conversion process. The octal number is first divided into digits according to place value. Each octal digit is then converted into an equivalent binary number using only three digits. The steps of this procedure are shown in Figure 12–6.

Converting a decimal number to an octal number is a process of repetitive division by the number 8. After the quotient has been determined, the remainder is brought down as the place value. When the quotient is even with no remainder, a 0 is transferred to the place position. Assume now that the number 4098_{10} is to be converted to an octal equivalent. The procedure for making this conversion is outlined in Figure 12–7.

Converting a binary number to an octal number is an important conversion process of digital circuits. Binary numbers are first processed at a very high speed. An output circuit could then accept this signal and convert it to an octal signal displayed on a readout device.

Assume that the number 110100100_2 is to be changed to an equivalent octal number. The digits must first be divided into groups of three, starting at the octal point. Each binary group is then converted into an

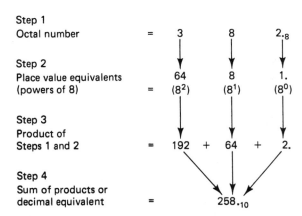

Step 1 Octal number	=	3	8	$2._8$
Step 2 Place value equivalents (powers of 8)	=	64 (8^2)	8 (8^1)	1. (8^0)
Step 3 Product of Steps 1 and 2	=	192 +	64 +	2.
Step 4 Sum of products or decimal equivalent	=		$258._{10}$	

FIGURE 12–5   Octal-to-decimal conversion.

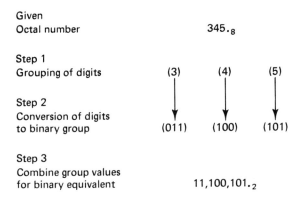

Given
Octal number   345.$_8$

Step 1
Grouping of digits   (3)   (4)   (5)

Step 2
Conversion of digits
to binary group   (011)   (100)   (101)

Step 3
Combine group values
for binary equivalent   11,100,101.$_2$

FIGURE 12–6   Octal-to-binary conversion.

equivalent octal number. These numbers are then combined, while remaining in their same respective places, to represent the equivalent octal number. The conversion steps are outlined in Figure 12–8.

HEXADECIMAL NUMBER SYSTEM

The *hexadecimal number system* is used in digital systems to process
large number values. The radix, or base, of this system is 16, which
means that the largest number used in a place is 15. Digits used by this
system are the numbers 0–9 and the letters A–F. The letters A–F are

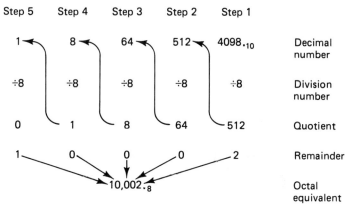

Step 5	Step 4	Step 3	Step 2	Step 1	
1	8	64	512	4098.$_{10}$	Decimal number
÷8	÷8	÷8	÷8	÷8	Division number
0	1	8	64	512	Quotient
1	0	0	0	2	Remainder
		10,002.$_8$			Octal equivalent

FIGURE 12–7   Decimal-to-octal conversion.

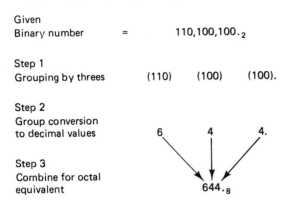

Given
Binary number   =   110,100,100.$_2$

Step 1
Grouping by threes   (110)   (100)   (100).

Step 2
Group conversion
to decimal values   6   4   4.

Step 3
Combine for octal
equivalent   644.$_8$

FIGURE 12–8   Binary-to-octal conversion.

used to denote the digits 10–15, respectively. The place value of digits to the left of the hexadecimal point are the powers of 16: $16^\circ = 1$, $16^1 = 16$, $16^2 = 256$, $16^3 = 4096$, $16^4 = 65,536$, etc.

The process of changing a hexadecimal number to a decimal number is achieved by procedures similar to those outlined for other conversions. Initially, a hexadecimal number is recorded in proper digital order as shown in Figure 12–9. The place values, or powers of the base, are then positioned under each respective digit in Step 2. The values of Steps 1 and 2 are then multiplied together to indicate place values. In hexadecimal conversions the value derived in Step 3 is usually added to simplify letter digit assignments. The values in Steps 2 and 3 are then multiplied together. Step 4 is the addition of these product values. Adding these values together in Step 5 gives the decimal equivalent value of a hexadecimal number.

The process of changing a hexadecimal number to a binary equivalent is a simple grouping operation. Figure 12–10 shows the steps for

Step 1:	Hexadecimal number	=	1	2	C	D.$_{10}$
Step 2:	Place value equivalents (powers of 16)	=	4096s	256s	16s	1s.
Step 3:	Place value digits	=	1	2	12	13.
Step 4:	Product of Steps 2 and 3	=	4096 + 512 + 192 + 13.			
Step 5:	Sum of products or decimal equivalent	=	4813.$_{10}$			

FIGURE 12–9   Hexadecimal-to-decimal conversion.

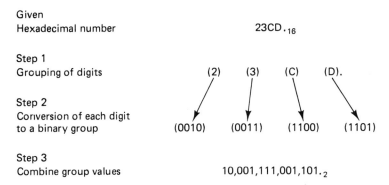

FIGURE 12–10   Hexadecimal-to-binary conversion.

making this conversion. Initially, the hexadecimal number is separated into digits in Step 1. Each digit is then converted to a binary number using four digits per group. Step 3 shows the binary group combined to form the equivalent binary number.

The conversion of a decimal number to a hexadecimal number is achieved by repetitive division as used with other number systems. In this procedure the division is by 16 and remainders can be as large as 15. Figure 12–11 shows the steps for this conversion.

Converting a binary number to a hexadecimal equivalent is the reverse of the hexadecimal to binary process. Figure 12–12 shows the steps of this procedure. Initially, the binary number is divided in groups of four digits, starting at the hexadecimal point. Each number group is

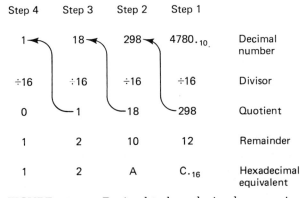

FIGURE 12–11   Decimal-to-hexadecimal conversion.

Given
Binary number   1,001,101,101,010.$_2$

Step 1
Grouping of fours   (0001)   (0011)   (0110)   (1010)

Step 2
Group conversion to   1   3   6   10
                             ↓   ↓   ↓   ↓
hexadecimal values   1   3   6   A

Step 3
Combine for
hexadecimal   136A.$_{16}$
equivalent

FIGURE 12–12   Binary-to-hexadecimal conversion.

then converted to a hexadecimal value and combined to form the hexadecimal equivalent number.

BINARY LOGIC CIRCUITS

In digital circuit design applications binary signals are far superior to those of the octal, decimal, or hexadecimal systems. Binary signals can be processed very easily through electronic circuitry since they can be represented by two stable states of operation. These states can be easily defined as on or off, 1 or 0, up or down, voltage or no voltage, right or left, or any other two-condition states. There must be no in-between state.

The symbols used to define the operational state of a binary system are very important. In positive binary logic the state of voltage, on, true, or a letter designation (such as A) is used to denote the 1 operational state. No voltage, off, false, or the letter $\overline{A}$ is commonly used to denote the 0 condition. A circuit can be set to either state and will remain until it is caused to change conditions.

Any electronic device that can be set in one of two operational states or conditions by an outside signal is said to be *bistable*. Relays, lamps, switches, transistors, diodes, and ICs may be used for this purpose. A bistable device has the capability of storing one binary digit or *bit* of information. By using many of these devices, it is possible to build an electronic circuit that will make decisions based upon the applied input signals. The output of this circuit is a decision based upon the operational conditions of the input. Since the application of bistable devices in digital circuits makes logical decisions, they are commonly called *binary logic circuits*.

Three basic circuits of this type are used to make simple logic decisions. These are the *AND* circuit, *OR* circuit, and the *NOT* circuit. The logic decision made by each circuit is different from the others.

Electronic circuits designed to perform logic functions are called gates. This term refers to the capability of a circuit to pass or block specific digital signals. An IF-THEN type of sentence is often used to describe the basic operation of a logic gate. For example, if the inputs applied to an AND gate are all 1, then the output will be 1. If a 1 is applied to any input of an OR gate, then the output will be 1. If an input is applied to a NOT gate, then the output will be the opposite.

AND GATES

An AND gate has two or more inputs and one output. If all inputs are in the 1 state simultaneously, then there will be a 1 at the output. Figure 12–13 shows a switch and lamp analogy of the AND gate, its sym-

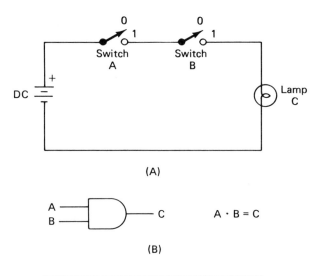

(A)

$A \cdot B = C$

(B)

Input switch A	Input switch B	Output lamp C
0	0	0
0	1	0
1	0	0
1	1	1

(C)

FIGURE 12–13 (A) AND gate circuit. (B) AND gate symbol. (C) AND gate truth table.

bol, and a truth table. In Figure 12–13A, when a switch is turned on, it represents a 1 condition and off represents a 0. The lamp displays the condition by being a 1 when it is on and 0 when turned off. The switches are labeled A and B and the output lamp is C.

The operation of a gate is simplified by describing the input-output relationships in a table. The table in Figure 12–13C shows the alternatives at the inputs and the corresponding outputs that occur as a result. This description of a gate is called a *truth table*. It shows the predictable operating conditions of a logic circuit.

Each input to an AND gate has two operational states of 1 and 0. A two-input AND gate has 2^2, or four, possible combinations that influence the output. A three-input gate has 2^3, or 8, combinations, while a four-input has 2^4, or 16, combinations. These combinations are normally placed in the truth table in binary progression order. For a two-input gate this is 00, 01, 10, and 11, which shows the binary count of 0, 1, 2, and 3 in order.

The AND gate of Figure 12–13A only produces a 1 output when switches A and B are both 1. Mathematically, this action is described as $A \cdot B = C$. This expression shows the multiplication operation. The symbol of an AND gate is shown in Figure 12–13B.

OR GATES

An OR gate has two or more inputs and one output. Like the AND gate, each input to the OR gate has two possible states: 1 or 0. The output of this gate produces a 1 when either or both inputs are 1. Figure 12–14 shows a lamp and switch analogy of the OR gate, its symbol, and a truth table.

An OR gate produces a 1 output when both switches are 1 or when either switch A or B is 1. Mathematically, this action is described as $A + B = C$. This expression shows OR addition. This gate is used to make logic decisions of whether or not a 1 appears at either input. The truth table of Figure 12–14C shows that if any input is a 1, the output will be a 1.

NOT GATES

A NOT gate has one input and one output. The output of a NOT gate is opposite to that of the input state. Figure 12–15 shows a switch and lamp NOT gate analogy, its symbol, and truth table. When the switch of Figure 12–15A is on (in the 1 state), it shorts out the lamp. Placing the switch in the off condition (0) causes the lamp to be on, or in the 1 state. NOT gates are also called *inverters*.

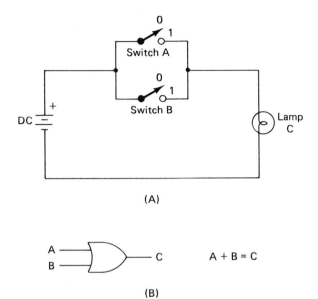

(A)

A + B = C

(B)

Input switch A	Input switch B	Output lamp C
0	0	0
0	1	1
1	0	1
1	1	1

(C)

FIGURE 12–14 (A) OR gate circuit. (B) OR gate symbol. (C) OR gate truth table.

COMBINATION LOGIC GATES

When a NOT gate is combined with an AND gate or an OR gate, it is called a combination logic gate. A NOT-AND gate is called a *NAND gate*. This gate is an inverted AND gate. Figure 12–16 shows a simple switch and lamp circuit analogy of the NAND gate, its symbol, and truth table.

The NAND gate is an inversion of the AND gate. When switches A and B are both on (in the 1 state), lamp C is off (0). When either or both

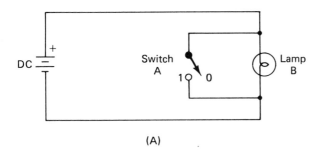

(A)

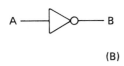

(B)

Input switch A	Output lamp B
0	1
1	0

(C)

FIGURE 12–15   (A) NOT gate circuit. (B) NOT gate symbol.
(C) NOT gate truth table.

switches are off, lamp C is in the on, or 1, state. Mathematically, the operation of a NAND gate is $A \cdot B = \overline{C}$. The bar over C denotes the inversion or negative function of the gate.

A combination NOT-OR, or NOR, gate produces a negation of the OR function. Figure 12–17 shows a switch and lamp circuit analogy of this gate, its symbol, and truth table. Mathematically, the operation of a NOR gate is $A + B = \overline{C}$. A 1 will only appear at the output when A is 0 and B is 0.

The logic gates discussed here illustrate basic gate operation. The switch and lamp input-output analogies are simple ways of showing gate operation. In actual digital electronic applications, solid-state components are ordinarily used to accomplish gate functions. Integrated circuits (ICs) which are used as logic gates are discussed in Chapter 13.

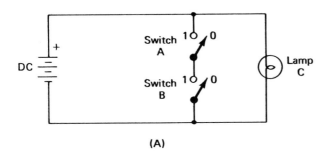

(A)

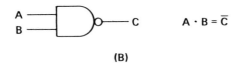

(B)

Input switch A	Input switch B	Output lamp C
0	0	1
0	1	1
1	0	1
1	1	0

(C)

FIGURE 12–16   (A) NAND gate circuit. (B) NAND gate symbol.
(C) NAND gate truth table.

BOOLEAN ALGEBRA

Boolean algebra is a special form of algebra that was designed to show the relationships of logic operations. This form of algebra is ideally suited for analysis and design of binary logic systems. Through the use of Boolean algebra, it is possible to write mathematical expressions that describe specific logic functions. Boolean expressions are more meaningful than complex word statements or elaborate truth tables. The laws that apply to Boolean algebra are used to simplify complex expressions. Through this type of operation, it may be possible to reduce the number

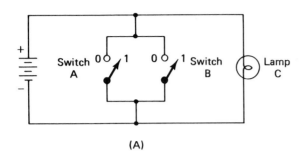

(A)

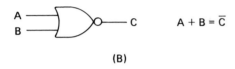

(B)

Input switch A	Input switch B	Output lamp C
0	0	1
0	1	0
1	0	0
1	1	0

(C)

FIGURE 12–17   (A) NOR gate circuit. (B) NOR gate symbol.
(C) NOR gate truth table.

of logic gates needed to achieve a specific function before the circuits
are designed.

In Boolean algebra the variables of an equation are assigned letters
of the alphabet. Each variable then exists in states of 1 or 0 according to
its condition. The 1, or true state, is normally represented by a single
letter such as A, B, or C. The opposite state or condition is then described as 0, or false, and is represented as $\overline{A}$ or A'. This is described
as NOT A, A negated, or A complemented.

Boolean algebra is somewhat different from conventional algebra
with respect to mathematical operations. The Boolean operations are expressed as follows:

Multiplication: A AND B, *AB*, *A · B*
OR Addition: A OR B, *A + B*
Negation or Complementing: NOT A, $\overline{A}$, *A'*

Assume that a digital logic circuit has three input variables, *A*, *B*, and *C*. The output circuit should operate when only *C* is on by itself, or when *A*, *B*, and *C* are all on at the same time.

Expressing this statement in a Boolean expression describes the desired output. Eight different combinations of *A*, *B*, and *C* exist in this expression because there are three inputs. (2^3). Only two of those combinations should cause a signal that will actuate the output. When a variable is not on (0), it is expressed as a negated letter. The original statement is expressed as follows: With *A*, *B*, and *C* on, or with *A* off, *B* off, and *C* on, an output (*X*) will occur:

$$ABC + \overline{AB}C = X$$

A truth table illustrates if this expression is achieved or not. Figure 12–18 shows a truth table for this equation. First, *ABC* is determined by multiplying the three inputs together. A 1 appears only when the *A*, *B*, and *C* inputs are all 1. Next the negated inputs *A* and *B* are determined. Then the product of inputs *C*, $\overline{A}$, and $\overline{B}$ are listed. The next column shows the addition of *ABC* and $\overline{AB}C$. The output of this equation shows that a 1 output is produced only when $\overline{AB}C$ is 1 or when *ABC* is 1.

A logic circuit to accomplish this Boolean expression is shown in Figure 12–19. Initially the equation is analyzed to determine its primary operational function. Step 1 shows the original equation. The primary function is addition, since it influences all parts of the equation in some way. Step 2 shows the primary function changed to a logic gate diagram.

$ABC + \overline{AB}C = X$

Inputs A B C	ABC	$\overline{A}$	$\overline{B}$	$\overline{AB}C$	Output $ABC + \overline{AB}C$
0 0 0	0	1	1	0	0
0 0 1	0	1	1	1	1
0 1 0	0	1	0	0	0
0 1 1	0	1	0	0	0
1 0 0	0	0	1	0	0
1 0 1	0	0	1	0	0
1 1 0	0	0	0	0	0
1 1 1	1	0	0	0	1

FIGURE 12–18 Truth table for Boolean equation.

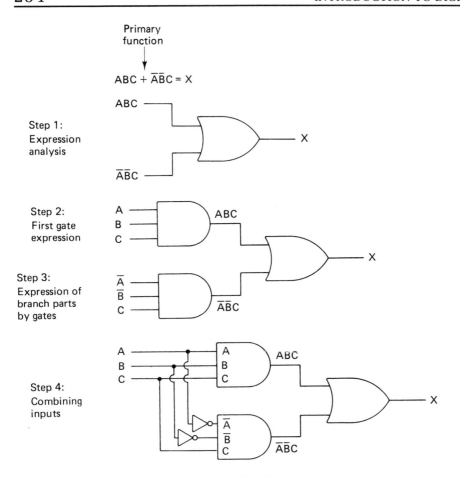

Step 1:
Expression
analysis

Step 2:
First gate
expression

Step 3:
Expression of
branch parts
by gates

Step 4:
Combining
inputs

FIGURE 12–19   Logic circuit to accomplish Boolean equation.

Step 3 shows the branch parts of the equation expressed by logic diagrams, with AND gates used to combine terms. Step 4 completes the process by connecting all inputs together. The circles at inputs $\overline{AB}$ of the lower AND gate are used to achieve the negation function of these branch parts.

The general rules for changing a Boolean equation into a logic circuit diagram are very similar to those outlined. Initially the original equation must be analyzed for its primary mathematical function. This is then changed into a gate diagram that is inputed by branch parts of the equation. Each branch operation is then analyzed and expressed in gate form. The process continues until all branches are completely expressed in diagram form. Common inputs are then connected together.

FLIP-FLOPS

Flip-flops are used as memory devices in digital circuits. They can be made to hold an output state even when the input is completely removed. They can also be made to change an output when an appropriate input signal occurs.

The *reset-set (R-S) flip-flop* is shown in Figure 12–20. The logic diagram symbol and truth table for this device are more complicated than that of a simple logic gate. They show the different states of the device before an input occurs and how they change after the input has arrived. Two of the operating conditions produce an unpredictable output. In this state the first input produces an output only by coincidence.

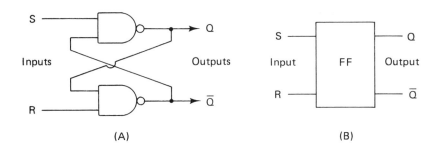

(A) (B)

Applied inputs		Previous outputs		Resulting outputs		
S	R	Q	$\overline{Q}$	Q	$\overline{Q}$	
0	0	1	0	1/0	0/1	Unpredictable
0	1	1	0	1	0	
1	0	1	0	0	1	
1	1	1	0	1	0	
0	0	0	1	0/1	1/0	Unpredictable
0	1	0	1	1	0	
1	0	0	1	0	1	
1	1	0	1	0	1	

(C)

FIGURE 12–20 R-S flip-flop. (A) Logic diagram with NAND gates. (B) Symbol. (C) Truth table.

In digital circuit applications flip-flops must often be set and cleared at specific times with respect to other operating circuits. This type of operation is achieved by manipulating flip-flops in step with a *clock pulse*. In this case the appropriate R-S inputs and clock pulse must all be present in order to cause a state change. A device of this type is called an R-S triggered flip-flop, or simply an *R-S-T flip-flop*.

The truth table of an R-S-T flip-flop is basically the same as that of an R-S flip-flop. It will only cause a state change when the clock pulse arrives at the T input. A two-input AND gate is simply added to the set and reset inputs to accomplish this operation. Figure 12–21 shows the R-S-T flip-flop circuits, its logic symbol, and truth table.

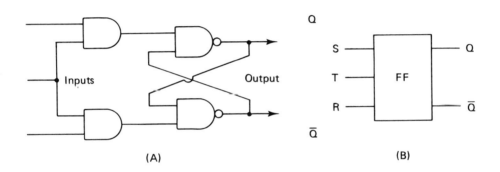

(A)   (B)

Applied inputs			Previous outputs		Resulting outputs		
S	R	T	Q	$\overline{Q}$	Q	$\overline{Q}$	
0	0	0	1	0	1/0	0/1	Unpredictable
0	1	1	1	0	1	0	
1	0	0	1	0	0	1	
1	1	1	1	0	1	0	
0	0	0	0	1	0/1	1/0	Unpredictable
0	1	1	0	1	1	0	
1	0	0	0	1	0	1	
1	1	1	0	1	0	1	

(C)

FIGURE 12–21   R-S-T flip-flop. (A) Logic diagram. (B) Symbol. (C) Truth table.

Another flip-flop circuit used in digital systems is the *JK flip-flop*. This circuit is somewhat unusual because it has no unpredictable output states. It can be set by applying a 1 to the *J* input and can be cleared by feeding a 1 to the *K* input. A 1 signal applied to both *J* and *K* inputs simultaneously causes the output to change states or toggle. A 0 applied to both inputs at the same time does not initiate a state change. The inputs of a JK flip-flop are controlled directly by clock pulses.

Figure 12–22 shows a logic diagram, symbol, and truth table for the JK flip-flop. There are no unpredictable output states for this device. Several modifications of the basic JK flip-flop are available today. These include those with preset and preclear inputs that are used to establish sequential operations at a precise time. Flip-flops are commonly used as the basic logic element for counting operations, temporary memory, and sequential switching operations.

DIGITAL COUNTERS

One of the most versatile and important logic devices of a digital system is the *counter*. Counters are used to count a wide variety of objects in a number of different digital system applications; however, they essentially count only one thing—electronic pulses. These pulses may be produced electronically by a clock mechanism, electromechanically, acoustically, or by a number of other processes.

BINARY COUNTERS

A common application of a digital counter is to count numerical information in binary form. This type of circuit uses flip-flops connected so that the Q output of the first circuit drives the trigger or clock input of the next circuit. Each flip-flop therefore has a divide-by-two function.

Figure 12–23 shows JK flip-flops (FFs) connected to achieve binary counting. The counter in Figure 12–23A is called a *binary ripple counter*. Each flip-flop in this circuit has the *J* and *K* inputs held at a logic 1 level. Each clock pulse applied to the input of FF_1 will then cause a change in state. The flip-flops only trigger on the negative-going part of the clock pulse. The output of FF_1 will therefore alternate between 1 and 0 with each pulse. A 1 output will appear at Q of FF_1 for every two input pulses. This means that each flip-flop has a divide-by-two function. Five flip-flops connected in this manner will produce a 2^5, or 32, count. The largest count in this case is 11111_2 (31_{10}). The next applied pulse clears the counter so that 0 appears at all the Q outputs.

By grouping three flip-flops together (Figure 12–23B), it is possible

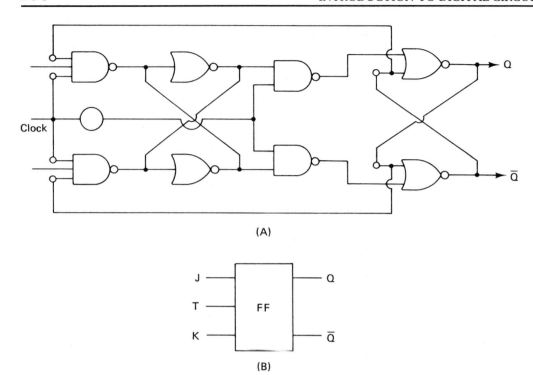

(A)

(B)

Applied inputs			Previous outputs		Resulting output		
J	K	T	Q	$\overline{Q}$	Q	$\overline{Q}$	
0	0	0	0	1	0	1	
0	1	1	0	1	0	1	
1	0	0	0	1	1	0	
1	1	1	0	1	1	0	← Toggle state
0	0	0	1	0	1	0	
0	1	1	1	0	1	0	
1	0	0	1	0	1	0	
1	1	1	1	0	0	1	

(C)

FIGURE 12–22   JK flip-flop. (A) Logic diagram. (B) Symbol.
(C) Truth table.

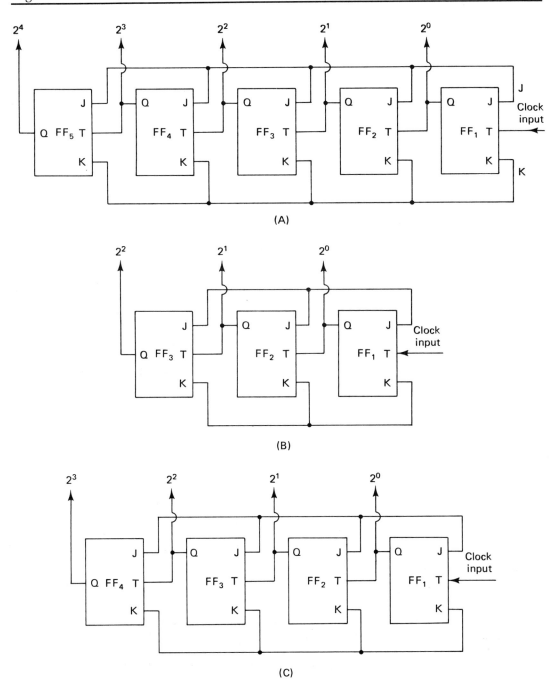

FIGURE 12–23   Digital counters using JK flip-flop. (A) Binary ripple counter (32 count). (B) Octal counter (8 count). (C) Hexadecimal counter (16 count).

to develop a binary-coded octal (BCO) counter. 111_2 is used to represent the seven count, or seven units of an octal counter. Two groups of three flip-flops connected in this manner would produce a maximum count of $111\text{-}111_2$, which represents 77_8, or 63_{10}.

By placing four flip-flops together in a group (Figure 12–23C), it is possible to develop the units part of a binary-coded hexadecimal (BCH) counter. Thus, 1111_2 is used to represent F_{16} or 15_{10}. Two groups of four flip-flops could be used to produce a maximum count of $1111\text{-}1111_2$, which would represent FF_{16} or 255_{10}. Each succeeding group of four flip-flops would be used to raise the counting possibility to the next power of 16.

DECADE COUNTERS

Since most of the mathematics that we use today is based upon the decimal (base 10) system, it is important to be able to count by this method. Digital systems are, however, designed to process information in binary form because of the ease with which a two-state signal can be manipulated. The output of a binary counter must therefore be changed into a decimal form before it can be used by one not familiar with binary numbers. The first step in this process is to change binary signals into a binary-coded decimal (BCD) form.

A *four-bit binary counter* is shown in Figure 12–24. In this counter 16 counts are achieved by the four flip-flops. To convert this counter

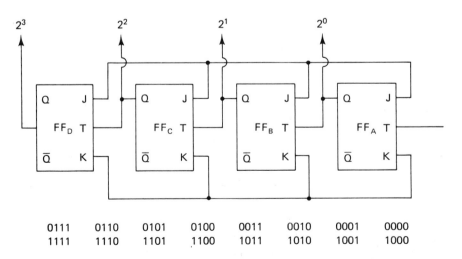

FIGURE 12–24   Four-bit binary counter.

into a decade counter, it must be caused to skip some of its counts. Notice the 16 counts listed below the binary counter, reading right to left.

A method of converting a binary counter into a decade counter is shown in Figure 12–25. In this method the first seven counts occur naturally as shown. Through these steps FF_D therefore remains at 1 during these counts. This is applied to the J input of FF_B, which permits it to trigger with each clock pulse.

At the seventh count, 1s appearing at the Q outputs of FF_B and FF_C are applied to the AND gate. This action produces a logic 1 and applies it to the J input of FF_D. Arrival of the next clock pulse triggers FF_A, FF_B, and FF_C into the off state and turns on FF_D. This represents the eight count.

When FF_D is in the on state, Q is 1 and $\overline{Q}$ is 0. This causes a 0 to be fed to the J input of FF_B, which now prevents it from triggering until cleared. Arrival of the next clock pulse cause FF_A to be set to a 1. This registers a 1001_2, which is the ninth count. Arrival of the next count clears FF_A and FF_D instantly. Since FF_B and FF_C were previously cleared by the seventh count, all 0s appear at the outputs. The counter has therefore cycled through the ninth count and returned to zero ready for the next input pulse.

Digital circuits of many types are now used in computer systems such as the one shown in Figure 12–26. This chapter is intended to be only an introduction to digital circuits. Answer the review questions, solve the problems, and perform the suggested laboratory activities to gain a better understanding of digital circuits.

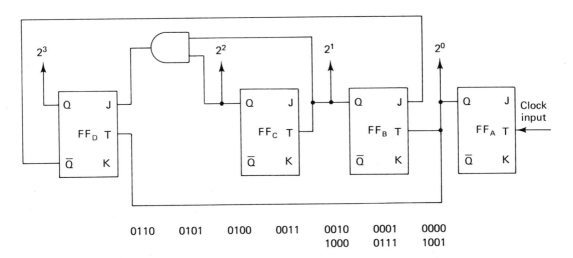

FIGURE 12–25   Decade or BCD counter.

FIGURE 12–26   Computer system. (Courtesy of TRW Resistive
Products Division)

REVIEW

1.   What is meant by the term *base*, or *radix*, of a number system?

2.   Discuss the binary number system.

3.   Discuss the BCD number system.

4.   Discuss the octal number system.

5.   Discuss the hexadecimal number system.

6.   What is meant by positive and negative logic in binary systems?

7.   What is a bistable device?

8.   Compare the input-output characteristics of two-input AND, OR, NAND, and NOR gates.

9.   What is a binary logic circuit?

10.   What is a truth table?

11.   What is Boolean algebra?

12.   What is the general procedure for changing a Boolean expression into a logic diagram?

13. What is a flip-flop?

14. Compare the R-S, R-S-T, and JK flip-flops.

15. How are flip-flops used to achieve binary counting?

16. Discuss binary counters.

17. Discuss decade counters.

PROBLEMS

1. How is the operation of the following gates expressed mathematically?

   a) AND

   b) OR

   c) NOT

   d) NAND

   e) NOR

2. Convert the following numbers to base 10 equivalents:

a) 11011_2	e) 131_8	i) 643_8
b) 110001_2	f) $11A_{16}$	j) $12A3_{16}$
c) 11000011_2	g) $149D_{16}$	k) $217C_{16}$
d) 23_8	h) 4062_8	l) $1011B_{16}$

3. Convert the following numbers to base 2:

a) 37_{10}	e) 105_8	i) 412_8
b) 61_{10}	f) $14A_{16}$	j) 1479_{16}
c) 123_{10}	g) $23D_{16}$	k) $10BA_{16}$
d) 127_8	h) 136_8	l) BDE_{16}

4. Develop a truth table and logic diagram for the expression $ABC + B = X$.

5. Develop a truth table and logic diagram for the expression $\overline{A}B + \overline{B}A = X$.

6. Express the following numbers in BCD form:

a) 137_{10} c) 195_{10}

b) 486_{10} d) 285_{10}

7. Change the following BCD expressions into decimal equivalents:

a) 0101-0111-1001

b) 1000-0110-0001

c) 1001-0010-0011

d) 1011-0111-1111

e) 0011-0010-0100

8. Convert the following decimal numbers to hexadecimal equivalents:

a) 4170_{10} e) 3159_{10}

b) 1210_{10} f) 5690_{10}

c) 1176_{10} g) 2137_{10}

d) 4153_{10} h) 583_{10}

9. Convert the following binary numbers to hexadecimal equivalents:

a) 10111110110_2

b) 100110111_2

c) 10110111100_2

d) 1101110111101_2

SUGGESTED LABORATORY ACTIVITIES

The following laboratory activities may be used to supplement the material discussed in this chapter.

1. Binary logic gates—Part 1.

a) Construct the two-input logic gate circuit shown in Figure 12–27. This circuit has an SN 7408 integrated circuit (IC) chip,

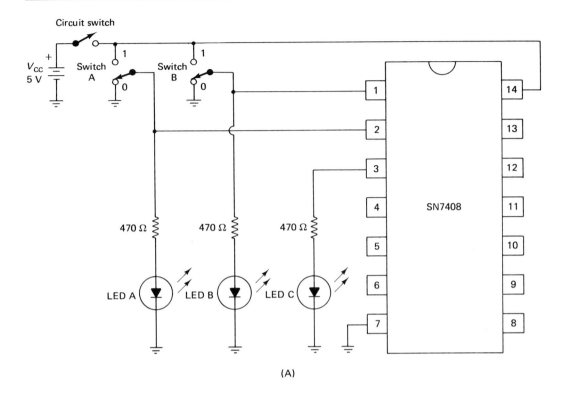

(A)

LED		
A	B	C
0	0	
0	1	
1	0	
1	1	

(B)

FIGURE 12–27   (A) Circuit. (B) Truth table.

which should be mounted in a 14-pin IC base, and three light-emitting diodes (LEDs) used to indicate input-output conditions.

b) Turn on the circuit switch and use switches A and B and the LED indicators to complete a truth table for this circuit.

c) What type of gate is this?

2. Binary logic gates—Part 2.

   a) Construct the three-input logic gate circuit shown in Figure 12–28 using an SN 7411 IC.

   b) Turn on the circuit switch and use switches A, B, and C and the LED indicators to complete a truth table for this circuit.

   c) What type of gate is this?

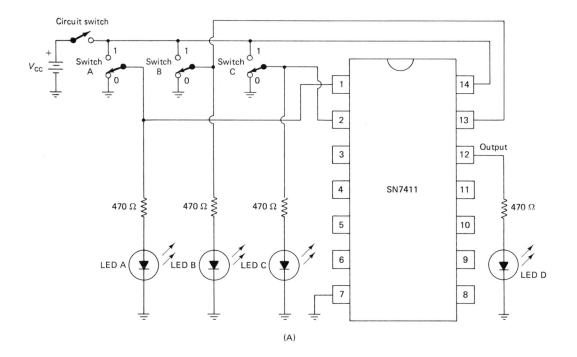

(A)

LED			
A	B	C	D
0	0	0	
0	0	1	
0	1	0	
0	1	1	
1	0	0	
1	0	1	
1	1	0	
1	1	1	

(B)

FIGURE 12–28   (A) Circuit. (B) Truth table.

3. Binary logic gates—Part 3.

   a) Construct the two-input logic gate circuit shown in Figure 12–29 using an SN 7432 IC.

   b) Turn on the circuit switch and use switches A and B and the LED indicators to complete a truth table for this circuit.

   c) What type of gate is this?

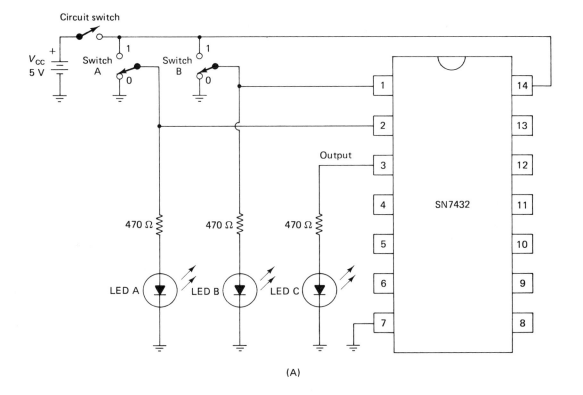

(A)

LED		
A	B	C
0	0	
1	0	
1	0	
1	1	

(B)

FIGURE 12–29   (A) Circuit. (B) Truth table.

4. Binary logic gates—Part 4.

   a) Construct the logic gate circuit shown in Figure 12–30 using an SN 7404 IC.

   b) Turn on the circuit switch and use switch *A* and the LED indicators to complete a truth table for this circuit.

   c) What type of gate is this?

Input	Output
LED A	LED B
0	
1	

(A)

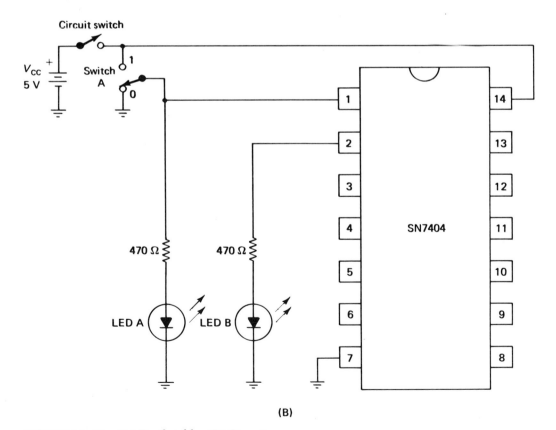

(B)

FIGURE 12–30   (A) Truth table. (B) Circuit.

5. Binary logic gates—Part 5.

   a) Construct the logic gate circuit shown in Figure 12–31 using an SN 7400 IC.

   b) Turn on the circuit switch and use switches *A* and *B* and the LED indicators to complete a truth table for this circuit.

   c) What type of gate is this?

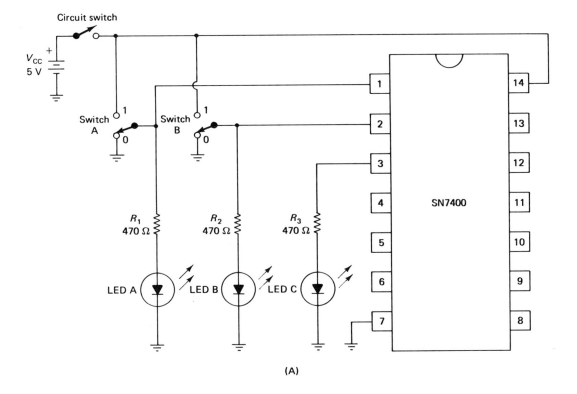

(A)

Gate 1		
Inputs		Output
A	B	C
0	0	
0	1	
1	0	
1	1	

(B)

FIGURE 12–31   (A) Circuit. (B) Truth table.

6. Binary logic gates—Part 6.

    a) Construct the logic gate circuit shown in Figure 12–32 using an SN 7402 IC.

    b) Turn on the circuit switch and use switches A and B and the LED indicators to complete a truth table for this circuit.

    c) What type of gate is this?

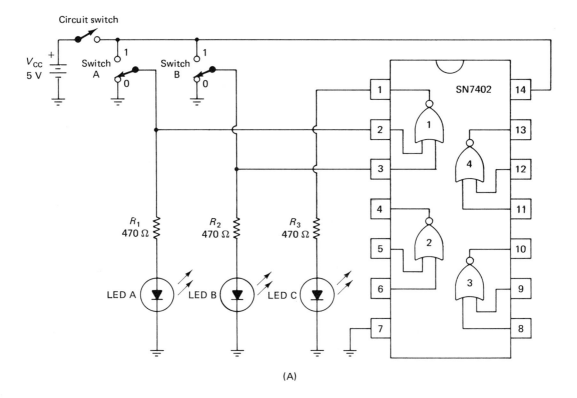

(A)

Gate 1		
Inputs		Output
A	B	C
0	0·	
0	1	
1	0	
1	1	

(B)

FIGURE 12–32   (A) Circuit. (B) Truth table.

7. R-S flip-flop circuit.

   a) Construct the R-S flip-flop circuit shown in Figure 12–33 using two of the NOR gates of the SN 7402 IC studied in laboratory activity 6.

   b) Turn on the circuit switch and use the R and S switches and the LED indicators to complete a truth table for this circuit.

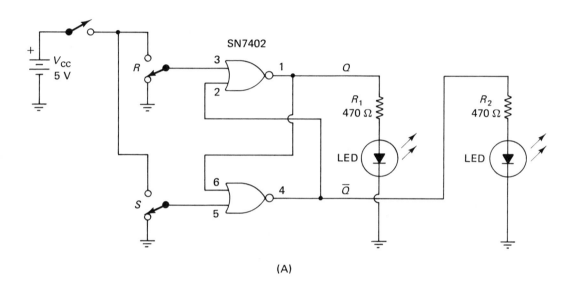

(A)

Inputs		Outputs	
R	S	Q	$\overline{Q}$
0	0		
0	1		
1	0		
1	1		

(B)

FIGURE 12–33   (A) Circuit. (B) Truth table.

8. R-S-T flip-flop circuit.

   a) Construct the R-S-T flip-flop circuit shown in Figure 12–34 using two of the NAND gates of the SN 7400 IC studied in laboratory activity 5.

b) Turn on the circuit switch and use the R, S, and T switches and the LED indicators to complete a truth table for this circuit.

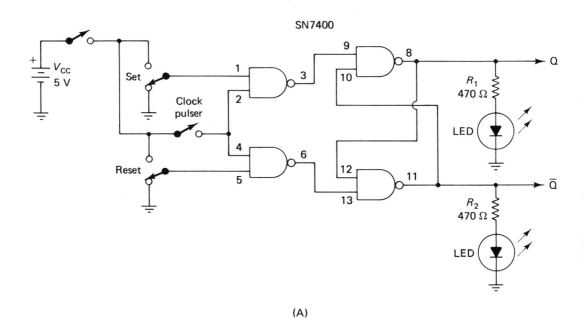

(A)

Set	Reset	Before clock		After clock	
		Q	$\overline{Q}$	Q	$\overline{Q}$
0	0	1	0		
0	1	1	0		
1	0	1	0		
1	1	1	0		
0	0	0	1		
0	1	0	1		
1	0	0	1		
1	1	0	1		

(B)

FIGURE 12–34   (A) Circuit. (B) Truth table.

CHAPTER 13
Introduction to Integrated Circuits and Operational Amplifiers

Integrated circuits (ICs) such as those shown in Figure 13–1 are very important electronic components. Within the last few years ICs have taken over many electronic circuit applications. Inside each small IC package are hundreds and even thousands of individual components. Transistors, resistors, diodes, and sometimes capacitors are built into IC chips. Transistors and FETs are key devices used in IC construction. IC chips may be designed to perform specific functions. Some very common types of ICs discussed in this chapter are *operational amplifiers (op-amps)* and *digital integrated circuits.*

An important feature of integrated circuits is their simplicity. These devices usually require an input of some type, an output connection, and a power source. The circuitry of these devices is self-contained in small packages. Figure 13–2 shows some common types of IC packages. The internal structure of ICs is very small and complex. The physical size of IC chips varies a great deal for different types. An *operational amplifier,* for example, has a chip size of approximately 0.003 × 0.003 in. Components within an IC chip cannot be changed or repaired when they fail to operate properly. The entire chip is removed from the circuit and replaced with a new one. Testing is easily done by observing

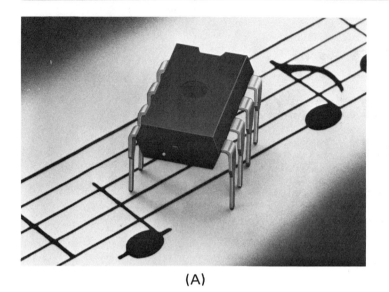

(A)

(B)

FIGURE 13–1 Integrated circuits. (A) Eight-pin dual in-line package (DIP) chip. (B) 28-pin metal-oxide semiconductor (MOS) IC used as a video pulse generator. (Courtesy of Siemens Corp.)

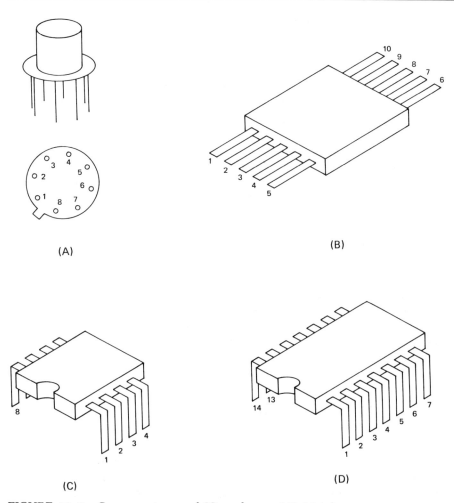

FIGURE 13–2 Common types of IC packages. (A) Metal can. (B) Flat-pack package. (C) Dual in-line (DIP) package. (D) 14-pin DIP package.

signals at the input and output terminals and checking power source voltage.

There are two basic types of ICs today: digital and linear. Digital ICs are essential parts of most digital systems today. They are used as gate circuits, flip-flops, memory systems, encoders, decoders, and many other digital applications.

Linear ICs usually perform some type of amplification. Two examples are operational amplifiers and differential amplifiers.

OPERATIONAL AMPLIFIERS

Many of the linear ICs used today are called operational amplifiers or op-amps. These devices have high-gain capabilities. Op-amps are used as signal amplifiers, waveshaping, impedance matching, and other applications. The symbol shown in Figure 13–3 is used for op-amps. A positive (+) sign on the input lead represents the *noninverting* input. The negative (−) sign indicates the *inverting* input. The internal structure of an op-amp IC has hundreds of components formed together. Electrical isolation among components is achieved by polarization of the material by using dielectric barriers. Metal plating is used to connect individual parts to external lead terminals. The entire assembly is then placed in an enclosure or package. The TO-5 package is similar to the one used for transistors and FETs. The *dual in-line package* (DIP) comes in many sizes and is very popular today.

Op-amps are high-gain amplifiers. They are designed to operate over a wide range of voltages. Their internal structure is small and complex, with components connected externally to determine their operating capabilities. Voltage gain (output-input ratio) can be 100,000 or more when the device is operated in an *open-loop* circuit. As a *voltage follower*, gain may be as low as 1.

DIFFERENTIAL AMPLIFIERS

The operation of op-amps is based upon differential amplifier circuits. A simplification of a differential amplifier circuit is shown in Figure 13–4. Transistors Q_1 and Q_2 form the basis of the differential amplifier circuit. Q_3 serves as constant current source for the amplifier. With no input signal applied to Q_1 or Q_2, no output appears. If identical values of input are applied to both Q_1 and Q_2, there is still no output. Each transistor

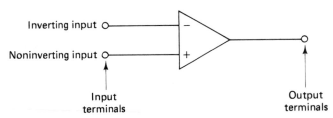

FIGURE 13–3   Symbol for op-amp.

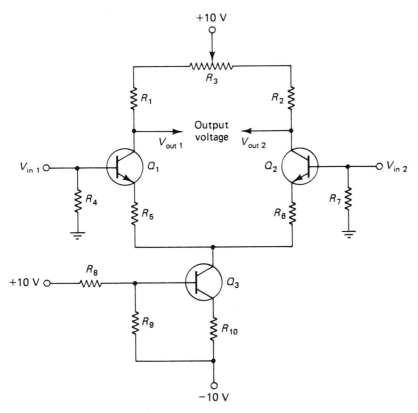

FIGURE 13–4   Differential amplifier circuit.

conducts the same amount and the output signal is balanced. There is no potential difference between the two outputs (V_{out1} and V_{out2}). No output signal appears when $V_{out1} = V_{out2}$.

When signals of different voltage values are applied to the inputs of Q_1 and Q_2, the output becomes unbalanced. The transistor with the largest signal will conduct the most current. The other transistor will conduct less current, so a difference in the two outputs will occur. The output then indicates the conduction difference between Q_1 and Q_2. The circuit is therefore called a *differential amplifier*.

When a signal is applied only to the *inverting input* of a differential amplifier, the output will be amplified and inverted. A signal applied only to the *noninverting* input is amplified but not inverted in phase or polarity. For the circuit to respond in this way, the input that is not used must be grounded. The resultant voltage output will then be the difference between the input value and ground potential.

OP-AMP CIRCUITS

A high-gain amplifier circuit increases the signal level so that very small input signals can be amplified by an op-amp. This circuit operates in conjunction with a differential amplifier inside an op-amp package. An *emitter-follower* stage comes after the high-gain amplifier. The emitter-follower allows the op-amp to have a low impedance output, which permits it to be connected to a variety of different output devices. The combination of these circuits makes an op-amp a rather complex circuit. Operation of an op-amp is usually controlled by the selection of external component values.

OPEN-LOOP CIRCUITS

Figure 13–5 shows a diagram of an open-loop op-amp circuit using an LM 741 op-amp with pin numbers shown. In this circuit configuration gain is controlled by the strength of the input signal. When the input signal voltage is zero, the output voltage is zero. When an input signal is applied, the output voltage will rise proportionally. Voltage gains of 100,000 or more are typical for this circuit. If the output voltage rises to the value of the source voltage, the op-amp is said to be *saturated*. An input signal of only a few millivolts is needed to cause the op-amp to reach saturation.

Applications of the op-amp in an open-loop configuration are not

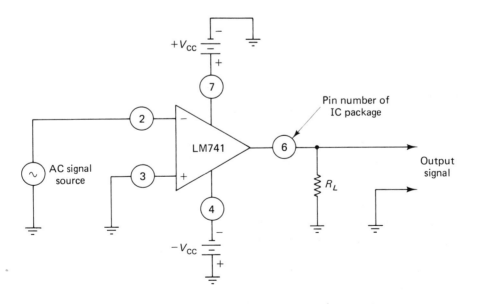

FIGURE 13–5   Open-loop op-amp circuit.

very common. Few circuits require a gain factor of 100,000. The input signal level for an open-loop circuit must be very small. Calculation of voltage amplification or gain is accomplished by the formula

$$\text{Voltage amplification} = \frac{\textbf{voltage output}}{\textbf{voltage input}} \quad \text{or} \quad A_v = \frac{V_o}{V_{in}}$$

The voltage amplification formula may be transposed to determine the input voltage: $V_{in} = V_o/A_v$.

Assume an A_v value of 50,000 for an open-loop op-amp circuit. The output voltage (V_o) cannot exceed the value of the source voltage. If the source voltage (V_{CC}) is 12 V, the input voltage value is calculated as

$$V_{in} = \frac{V_o}{A_v} = \frac{12}{50,000} = 0.00024 \text{ V} \quad \text{or} \quad 0.24 \text{ mV}$$

Saturation occurs when the input voltage reaches 0.24 mV. An AC signal greater than 0.24 mV peak to peak (p-p) would cause distortion of the output. An op-amp in an open-loop circuit is very easy to distort.

CLOSED-LOOP CIRCUITS

An op-amp used in a closed-loop circuit can provide voltage gain which can be adjusted to the desired value. In this circuit a feedback resistor network is connected between the output and input. Figure 13–6 shows an op-amp connected in a noninverting circuit. The output is of the same polarity as the input. The two inputs are balanced when resistors R_1 and R_2 are equal. R_2 is a calculated value based upon the parallel arrangement of R_1 and the feedback resistor (R_F). Gain of a noninverting closed-loop circuit is calculated by the formula

$$\text{Voltage amplification} = \frac{\textbf{input resistance} + \textbf{feedback resistance}}{\textbf{input resistance}}$$

or

$$A_v = \frac{R_i + R_F}{R_i}$$

Figure 13–7 shows an inverting closed-loop op-amp circuit. The input·signal is applied to the inverting $(-)$ terminal. The feedback network is composed of R_1 and R_F. R_2 connects the noninverting input to ground. The value of R_2 is determined by the parallel value of R_1 and R_F.

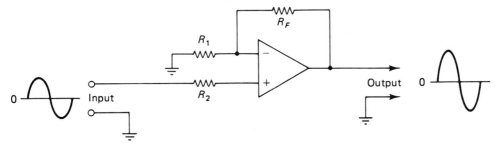

FIGURE 13–6   Closed-loop noninverting op-amp circuit.

R_2 is usually smaller than R_1. The voltage gain of inverting a closed-loop amplifier circuit is calculated by the formula

$$\text{Voltage amplification} = \frac{\textbf{feedback resistance}}{\textbf{input resistance}}$$

or

$$A_v = \frac{R_F}{R_i}$$

Figure 13–8 shows a functional op-amp circuit connected in a non-inverting configuration. The numbers near each lead are the IC pin numbers. The LM741 is ordinarily housed in an eight-pin dual in-line package. The voltage sources are supplied from split power supplies such as those discussed in Chapter 11. Common leads of each side of the supply are connected to ground. The ground is connected to several points on the circuit. The gain of this circuit is determined by the calculated values of the feedback network.

A functional inverting op-amp circuit is shown in Figure 13–9. This amplifier has gain capabilities similar to the noninverting circuit.

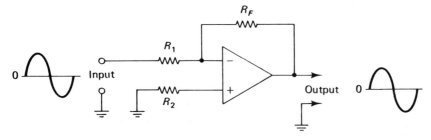

FIGURE 13–7   Closed-loop inverting op-amp circuit.

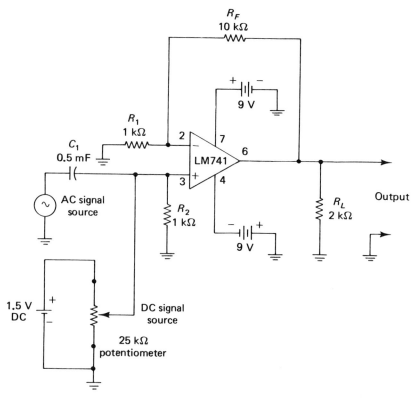

FIGURE 13–8   Noninverting op-amp circuit.

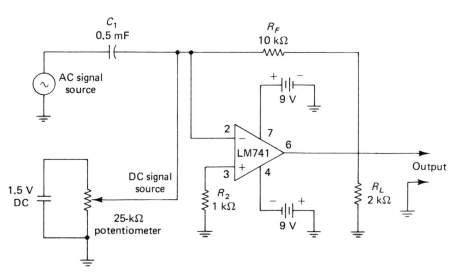

FIGURE 13–9   Inverting op-amp circuit.

The primary differences are the connections of the input circuit components.

IC AMPLIFIERS

Several companies manufacture IC amplifier chips. Some of these circuits are designed to perform one specific function. These functions include preamplification, linear signal amplification, and power amplification. Some ICs perform as a complete amplification system. An electric power source and a few external components are needed for operation. Low-power audio amplifiers and stereo amplifiers are examples of these ICs.

IC power amplifiers are often used in sound amplification systems. An IC audio power amplifier is shown in Figure 13–10. An internal circuit diagram of the IC is shown in Figure 13–10A, while a pin diagram is shown in part B. This particular chip has a differential amplifier input. Darlington transistor pairs are used to increase the input impedances for the inverting and noninverting inputs. When a signal is applied to one input, the other is ordinarily grounded. This IC responds to a difference in input signal levels, has a fixed gain of 20, and variable gain capabilities of up to 200. Without any external components connected to pins 1 and 8, the device has a gain of 20. A 10-mF capacitor connected between pins 1 and 8 causes a gain of 200. The output of the unit is a *Darlington transistor power amplifier*.

A circuit using the IC power amplifier is shown in Figure 13–11. Pins 1 and 8 have a capacitor-resistor combination which causes the gain to be less than 200. R_1 is a volume control for the circuit. The amplitude of the input signal is controlled by adjusting R_1. Applications of this IC include portable AM-FM radio amplifiers, tape player amplifiers, and television audio amplifiers.

THE NE/SE 555 IC

The NE/SE 555 IC is widely used today. This device uses a minimum of external components and a power source. Circuit design is easy to accomplish and operation is very reliable. This specific IC chip is available through a number of manufacturers. The number 555 usually appears in the manufacturer's part identification number (e.g., SN72555, MC14555, SE555, and LM555).

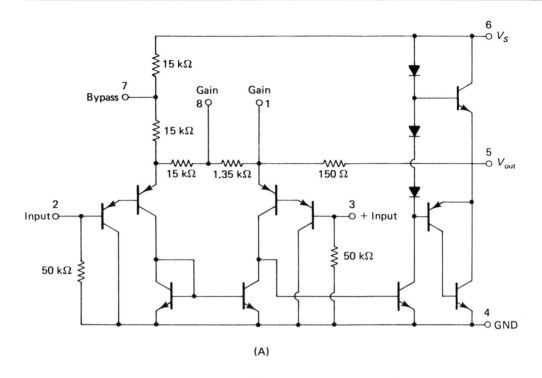

(A)

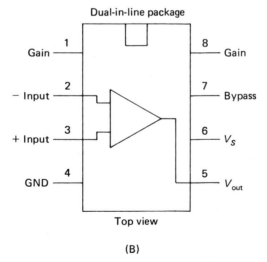

Top view

(B)

FIGURE 13–10 Integrated circuit audio power amplifier. (A) Internal circuit. (B) Pin diagram. (Courtesy of National Semiconductor Corp.)

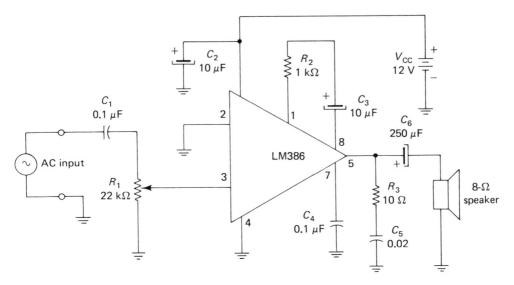

FIGURE 13–11   IC power amplifier circuit. (Courtesy of National Semiconductor Corp.)

The internal circuitry of a 555 IC may be viewed in functional blocks. The chip has two comparators, a bistable flip-flop, a voltage divider network, a discharge transistor, and an output stage. Figure 13–12 shows the functional blocks of a 555 IC. The voltage divider network of the IC consists of three 5-kΩ resistors. This network is connected across the power source ($+V_{CC}$) and ground. Voltage across the lower resistor is one-third of the source voltage. The point at pin 5 is two-thirds of the V_{CC} value. Pin 5 is the control voltage of the 555 IC.

The two *comparators* of the 555 IC act as an amplifying switch circuit. A reference voltage is applied to one input of each comparator. A voltage value applied to the other input causes a change in output when it is different from the reference value. Comparator 1 is referenced at a voltage of two-thirds of V_{CC} at its negative input. The other input is at pin number 6, which is called the *threshold* terminal. When the threshold voltage rises above two-thirds of V_{CC}, the output of the comparator is a positive value. This voltage is applied to the reset input of the flip-flop circuit.

Comparator 2 is referenced at a voltage of one-third of V_{CC}. The positive input of comparator 2 is connected to the lower voltage divider resistor. The *trigger* input is applied to the negative terminal of comparator 2. If the voltage of the trigger drops below one-third of V_{CC}, the comparator output becomes a positive value. The comparator output is applied to the set input of the flip-flop circuit.

The flip-flop of the 555 IC has reset and set inputs and one output. When the reset input is positive, the output is positive. A positive volt-

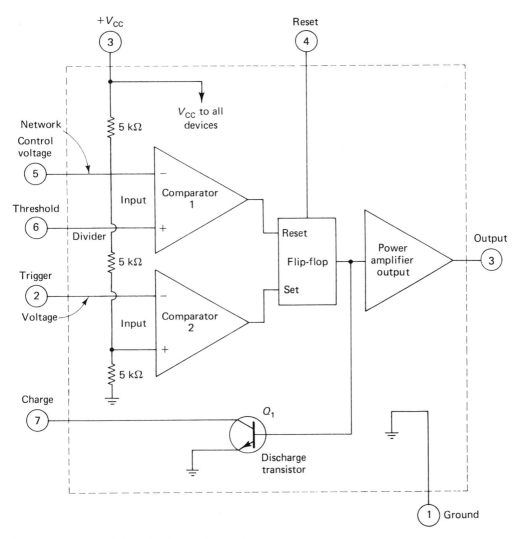

FIGURE 13–12   LM555 IC internal circuitry.

age to the set input causes the output to become negative. The output of the flip-flop depends upon the operational conditions of the two comparator inputs. The output of the flip-flop is applied to the output terminal (pin 3) and to the discharge transistor (Q_1). The discharge transistor is connected to pin 7, which is called the charge terminal. The output stage has a power amplifier and a signal inverter. A load circuit connected to the output (pin 3) will see either $+V_{CC}$ or ground, depending upon the value of the input signal. The output terminal switches between these two values. Load current values of up to 200 mA can be controlled by the output terminal. A load circuit connected to $+V_{CC}$

would be energized when pin 3 is at ground potential. When the output is at $+V_{CC}$ potential, the output will be off. A load circuit connected to ground will turn on when the output is at $+V_{CC}$ potential and off when the output is at ground potential. The output switches back and forth between these two states.

The output of the flip-flop is applied to the base of the discharge transistor (Q_1). When the flip-flop is reset (positive), it forward biases Q_1. The charge terminal (pin 7) connects to ground through Q_1. This causes pin 7 to be grounded. When the flip-flop is set (negative), it reverse biases Q_1. This causes pin 7 to be of infinite resistance. The charge terminal therefore has two states, shorted to ground or open with respect to ground.

APPLICATION OF THE 555 IC

A common application of the 555 IC is as a *time base generator* for clock or timing circuits and computers. This circuit configuration is called an *astable multivibrator*. The circuit is actually an *RC* oscillator whose waveshape and frequency are determined by an *RC* network.

The 555 IC astable multivibrator circuit shown in Figure 13–13 requires two resistors, a capacitor, and a power source. The output of the

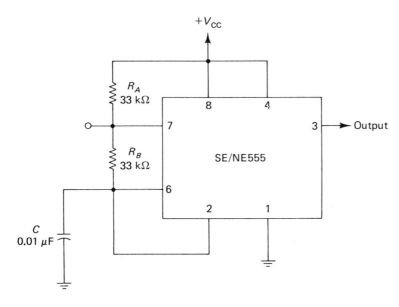

FIGURE 13–13   555 IC astable multivibrator circuit.

circuit is connected through pin 3. Pin 8 connects to $+V_{CC}$ and pin 1 to ground. The supply voltage (V_{CC}) can be from 5 to 15 V direct current. Resistor R_A is connected between $+V_{CC}$ and the discharge terminal (pin 7). Resistor R_B is connected between pins 7 and the threshold terminal (pin 6). A capacitor is connected between the threshold and ground. The trigger (pin 2) and threshold (pin 6) are connected together.

When power is first applied, the capacitor will charge through R_A and R_B. When the voltage at pin 6 (threshold) rises slightly above two-thirds of V_{CC}, it changes the state of comparator 1. This resets the flip-flop and causes its output to be positive. The output (pin 3) is then grounded and the base of Q_1 is forward biased. Q_1 then causes the capacitor *(C)* to discharge through R_B to ground.

When the charge voltage of the capacitor drops slightly below one-third of V_{CC}, it energizes comparator 2. Comparator 2 causes a positive voltage to go to the set input of the flip-flop, which sets the flip-flop. This causes the output of the flip-flop to be negative. The output (pin 3) then goes to $+V_{CC}$ potential and causes the base of Q_1 to be reverse biased. Since the charge terminal (pin 7) is open, *C* begins to charge again. *C* will charge to V_{CC} potential through R_A and R_B. The process is repeated from this point. The charge value of *C* varies between one-third and two-thirds of V_{CC}.

The resulting waveforms of the circuit are shown in Figure 13–14. The output frequency of the astable multivibrator is represented as $f = 1/T$. This frequency depends upon the total time needed to charge and discharge *C*. The charge time is represented by spaces t_1 and t_3. In seconds, t_1 is $0.693(R_A + R_B) \times C$. The discharge time is shown as t_2 and t_4 and is equal to $0.693(R_B \times C)$ seconds. The combined time for one oper-

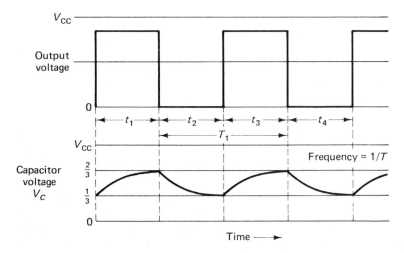

FIGURE 13–14   Astable multivibrator waveforms.

ational cycle is therefore $T = t_1 + t_2$ or $t_3 + t_4$. Combining t_1 and t_2 or t_3 and t_4 gives the frequency formula

$$f = \frac{1}{T} \quad \text{or} \quad \frac{1.44}{(R_A + 2R_B) \times C}$$

The resistance ratio of R_A and R_B is critical in the operation of an astable multivibrator. If R_B is more than half the value of R_A, the circuit astable multivibrator. If R_B is more than half the value of R_A, the circuit will not oscillate. This would prevent the trigger from dropping in value from two-thirds of V_{CC} to one-third of V_{CC}, so that the IC would not be able to trigger itself. Most manufacturers of ICs provide data charts that assist the user in selecting the correct R_A and R_B values with respect to capacitor values.

DIGITAL ICs

There are many applications of ICs in digital computer systems. Some of these applications, such as gate circuits, flip-flops, and counters, were discussed in Chapter 12.

LOGIC GATE CIRCUITS

The switch-lamp analogy of different gate circuits used in Chapter 12 showed basic logic operating characteristics in simplified form. In actual circuit applications logic gates have little resemblance to this type of circuit construction. Transistors, diodes, and other components are commonly connected together to achieve gate functions. The switching action of a gate is achieved very quickly by applying either forward or reverse biasing to solid-state components.

At one time logic circuits of the discrete component type were very popular. Developments in IC technology have now brought about a tremendous reduction in applications of discrete-component logic circuits. Multiple-gate structures are now being built inexpensively on single IC chips. Through these devices it is possible to construct complex logic circuits by simply interconnecting different logic gates. An understanding of basic logic functions is therefore much more important today than it has ever been. The truth table of a specific gate and its logic symbol terminal connections seem to be the two most important items needed in selecting an IC to perform a particular function. Figure 13–15 shows

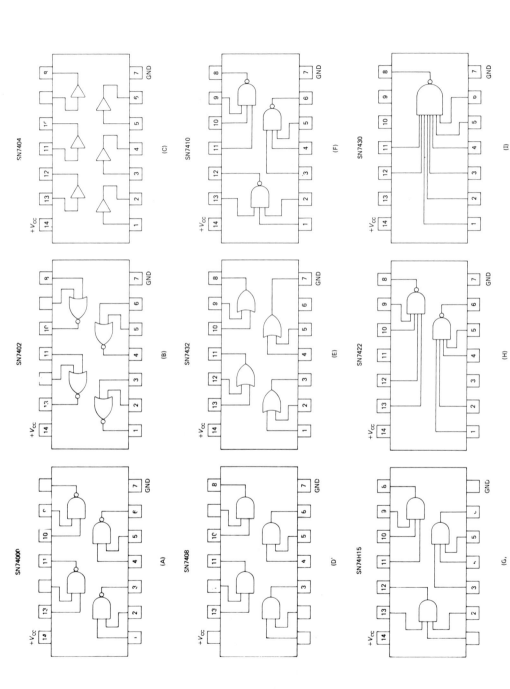

FIGURE 13–15 ICs used as digital logic gates. (A) SN7400: Quad NAND gate. (B) SN7402: Quad NOR gate. (C) SN7404: Hex inverter. (D) SN7408: Quad AND gate. (E) SN7432: Quad OR gate. (F) SN7410: Three 3-input NAND gates. (G) SN74H15: Three 3-input AND gates. (H) SN7422: Dual quad input NAND gates. (I) SN7430: 8-input NAND gate.

some typical pin connection diagrams of transistor-transistor logic (TTL) ICs used as logic gate circuits.

BINARY COUNTER CIRCUITS

Binary counters that contain four interconnected flip-flops are commonly built on one IC chip. Figure 13–16 shows the logic connections of a 4-bit binary counter. When used as a 4-bit counter, the FF_A output must be externally connected to the input of FF_B. Clock pulses applied to the input of FF_A produce a maximum count of $1111._2$ or $15._{10}$. Disconnecting FF_A from FF_B and applying the clock pulse to the FF_B input produces a 3-bit or BCO counter. The outputs of the flip-flops FF_A through FF_D are labeled A, B, C, and D, respectively.

Figure 13–17 shows an IC BCD counter. The operation of this IC is similar to the one just described. When this IC is used as a BCD counter,

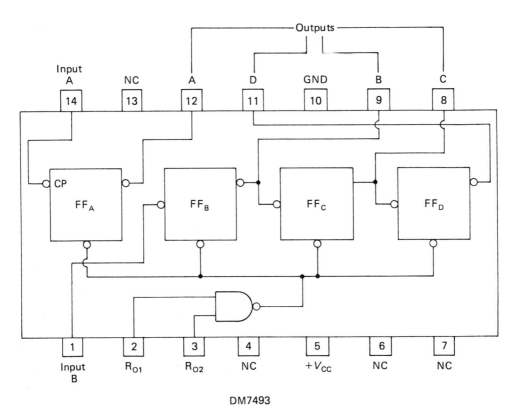

DM7493

FIGURE 13–16   IC used as a 4-bit binary counter.

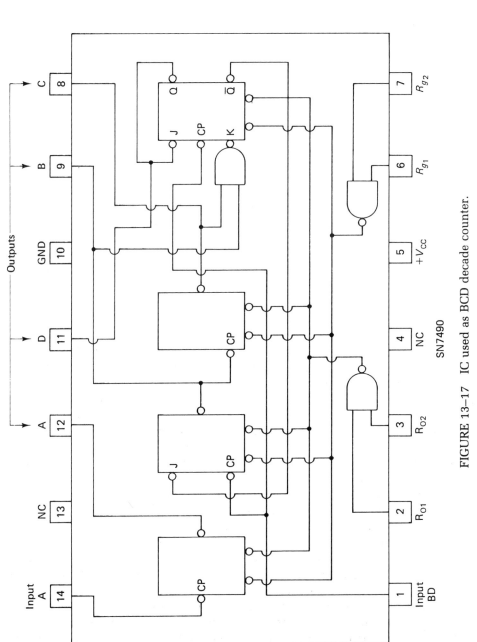

FIGURE 13–17 IC used as BCD decade counter.

the output of FF_A must be connected to the *BD* input. Omitting this connection and applying the clock pulses to the *BD* input produces a five count. With this option the counter is more versatile. The two NAND gates of this IC are used to set or reset the four flip-flops from an outside source. This counter triggers only on the negative-going part of clock pulse.

Figure 13–18 shows a BCD counter constructed from an SN7409 IC chip. The *ABCD* output display on the LEDs is the binary code for numbers 0–9. The input is a square-wave signal or clock pulse applied to pin 4. The output of a BCD counter is usually applied to a decoding device. The decoder changes binary data into an output that drives a *display device*.

Counters must change their output to a more practical method of display since it is difficult to read binary number displays. Display devices of several types have been developed. One of the most used number displays is called a *seven-segment display*. Calculators, watches, and electronic instruments use this type of display. A seven-segment display divides the decimal numbers 0–9 into lines. Four vertical lines and three horizontal lines are used to display numbers, with each segment illuminated separately. Figure 13–19 shows a 14-pin seven-segment display which plugs into an IC socket. Each segment may contain several LEDs in a line. The size of the display determines the number of LEDs in a line. All of the LEDs in a segment are energized at the same time.

Individual lines of a seven-segment display are labeled *a*, *b*, *c*, *d*, *e*, *f*, and *g*. A 7-bit binary count is applied to the display, with each segment fed by a part of the binary number. In a common-anode type of display, when the input is 0 (ground potential), the segment lights. In a common-cathode display, an input of 1 (+5 V) illuminates a segment.

Liquid crystal displays (LCDs) are another type of display device. This method of display is achieved by applying voltage to discrete bars of phosphorized silicon. When voltage is applied, the crystal material will change from a transparent state to an opaque condition that reflects ambient light. In a strict sense this action takes place when the phosphors are bombarded by electrons from the voltage source.

Liquid crystal digital devices are typical of the seven-segment type. The circuit construction is very similar to that of the seven-segment LED when energized. As a general rule, LCDs are commonly used in wristwatches, pocket calculators, and portable digital equipment. Very small amounts of electric energy are needed to produce a readout display. This is the LCD's primary advantage.

LCDs are primarily designed to reflect normal room light when they are energized. This type of display therefore develops significantly less light intensity when compared with other displays. Liquid crystal units are rarely ever used in industrial applications today because of this disadvantage.

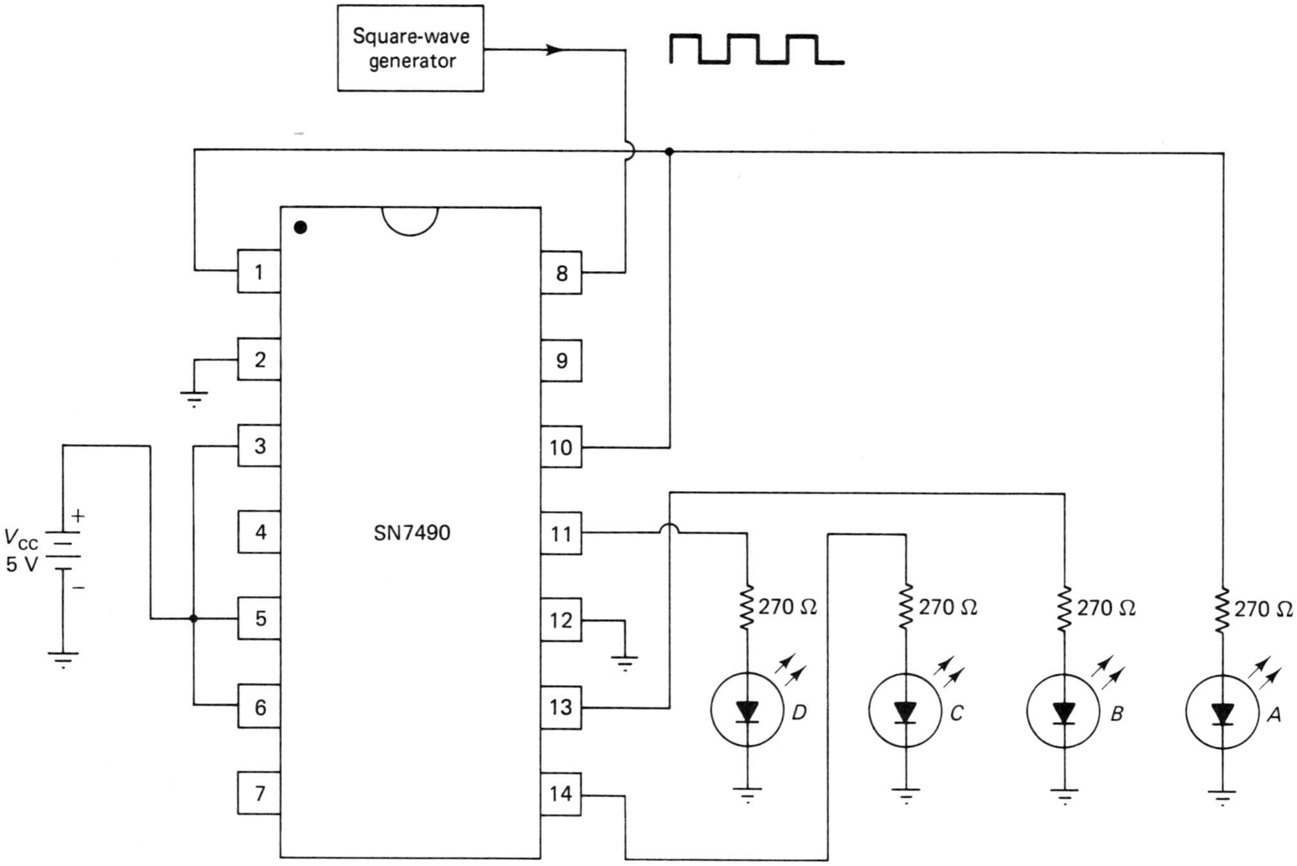

FIGURE 13-18 IC BCD counter circuit.

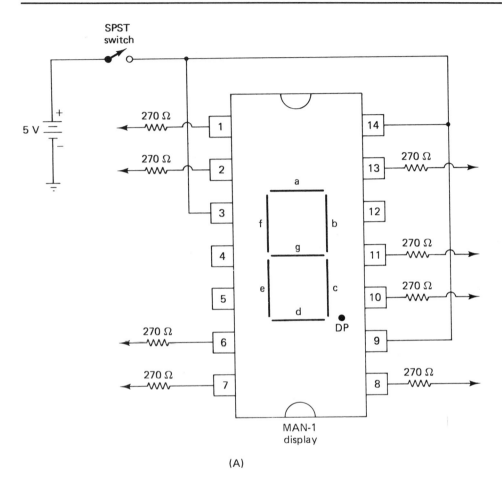

(A)

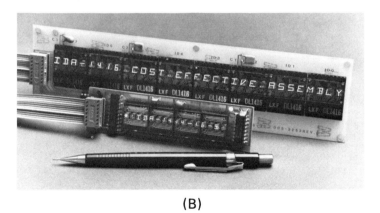

(B)

FIGURE 13–19 (A) 14-pin seven-segment display (DP—decimal point). (B) Seven-segment display assemblies for microprocessors (Courtesy of Siemens Corp.).

BCD inputs				7-segment outputs						
A	B	C	D	a	b	c	d	e	f	g
0	0	0	0	0	0	0	0	0	0	1
0	0	0	1	1	0	0	1	1	1	1
0	0	1	0	0	0	1	0	0	1	0
0	0	1	1	0	0	0	0	1	1	0
0	1	0	0	1	0	0	1	1	0	0
0	1	0	1	0	1	0	0	1	0	0
0	1	1	0	0	1	0	0	0	0	0
0	1	1	1	0	0	0	1	1	1	1
1	0	0	0	0	0	0	0	0	0	0
1	0	0	1	0	0	0	0	1	0	0

"0" output = grounded element
"1" output = open circuit

FIGURE 13–20   Truth table for seven-segment display.

DECODING CIRCUITS

Digital signals are usually in binary form. The input of a seven-segment display requires 7 bits of data. The number 4, for example, is 1001100 for the a–g segments. A truth table for a seven-segment number display is shown in Figure 13–20. The outputs do not correspond directly to the output of a counter circuit. A counter circuit usually provides binary data in a specific sequence. A BCD counter output for the number 4 is 0100. This data must be changed into the seven-segment number code to cause a 4 to appear on the display. The process of changing data of one form into another form is called *decoding*.

An IC used as a BCD-to-seven-segment *decoder-driver* is shown in Figure 13–21. The BCD input is applied to terminals A, B, C, and D (7, 1, 2, and 6). The seven-segment output appears at a, b, c, d, e, f, and g (13, 12, 11, 10, 9, 15, and 14). The IC also has a lamp test input, blanking input, and ripple blanking input. A 0 applied to the lamp test causes all segments of the display to be illuminated. A 0 applied to the blanking input turns off all segments.

For driving a common-anode display the outputs of the decoder must be 0 (grounded) to light a segment. Common-cathode displays must

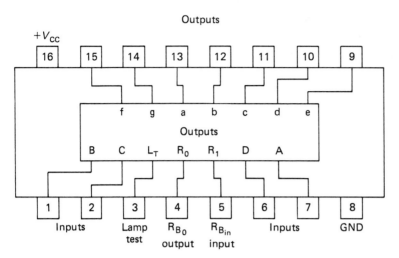

FIGURE 13–21   BCD to seven-segment decoder-driver IC.

be 1 (+V_{CC} potential) to light a segment. The decoder-driver chip used for a specific application depends upon the type of display employed. The truth table of Figure 13–20 shows that a 0 indicates the low or grounded output and the 1 output is an open circuit. The decoder of Figure 13–21 is used to drive a common-anode display. The number 0, for example, would illuminate all segments except g.

A *counting-timer circuit* which uses three ICs and a digital display is shown in Figure 13–22. This circuit operates as a decoder-driver for a seven-segment display. The 555 IC is used as a square-wave generator which can be adjusted to produce the desired clock pulse. The display counts 0–9 at a speed based on the pulse rate. R_1 is adjusted to determine the pulse rate. For a 1-s pulse rate, the capacitor (C) should be a value of 2 µF. The 0–9 counter can be extended to a 0–99 counter by adding an additional BCD counter, decoder, and display. Pin 11 of the units display is connected to the input (pin 14) of the tens display. The commons of each display unit must also be connected together. Each additional counter would be connected in the same manner. The count would be 0, 1, 2, 3, 4, 5, 6, 7, 8, 9, 10, 11, and continue up to 99. The next count above 99 would clear the counter to 00 and the sequence would start over. This circuit is one of many IC applications in digital systems.

In this chapter an introduction to ICs and operational amplifiers has been provided. Answer the following review questions, solve the problems, and perform the suggested laboratory activities to supplement the understanding of the material in this chapter.

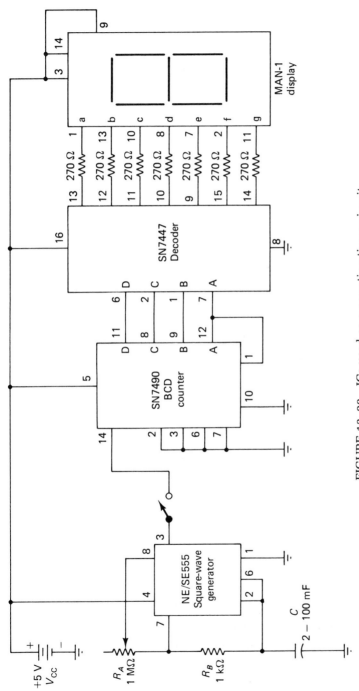

FIGURE 13-22   ICs used as a counting-timer circuit.

REVIEW

1. What is an integrated circuit chip?

2. What is a digital IC?

3. What is a linear IC?

4. What is an operational amplifier?

5. What is meant by inverting and noninverting op-amp inputs?

6. What are some types of packages used for ICs?

7. What is a differential amplifier?

8. What is an open-loop op-amp circuit?

9. What is a closed-loop op-amp circuit?

10. How is voltage amplification calculated?

11. Discuss the NE/SE 555 IC.

12. Discuss the use of the NE/SE 555 IC as an astable multivibrator.

13. How is charge time of the NE/SE 555 IC calculated?

14. How is discharge time of the NE/SE 555 IC calculated?

15. How is frequency of the NE/SE 555 IC calculated?

16. What are some digital applications of ICs?

17. What is a seven-segment display?

18. For a common-cathode seven-segment display, which segments are at $+V_{CC}$ potential when the number 6 is displayed?

19. What is decoding?

20. Make a sketch of a counting-timer circuit which could be used to count to 99.

PROBLEMS

1. An amplifier circuit has an input of 0.02 V and an output of 8 V. What is the voltage amplification (V_A)?

2. An op-amp circuit has an input of 0.05 mW and an output of 0.6 V. What is the V_A?

3. An open-loop op-amp circuit has an V_A of 100,000 and a V_{CC} of 10 V. What is the maximum input voltage?

4. An open-loop op-amp circuit has an V_A of 70,000 and a V_{CC} of 12 V. What is the maximum input voltage?

5. A noninverting closed-loop op-amp circuit has an input resistance (R_i) of 1 kΩ and a feedback resistance (R_F) of 99 kΩ. What is the voltage amplification (V_A)?

6. For the circuit of problem 5, what is the V_A if R_i is changed to 9.9 kΩ and R_F to 100 Ω?

7. An inverting closed-loop op-amp circuit has a feedback resistance of 100 kΩ and an input resistance of 100 Ω. What is its V_A?

8. For the circuit of problem 7, what is the V_A if R_i is changed to 200 kΩ and R_F to 1 kΩ?

9. What is the charge time of an NE/SE 555 circuit with the following values: $R_A = 100$ kΩ, $R_B = 50$ kΩ, $C = 20$ μF?

10. What is the discharge time of the circuit of problem 9?

11. What is the frequency value of the circuit of problem 9?

12. What is the charge time of an NE/SE 555 circuit with the following values: $R_A = 50$ kΩ, $R_B = 50$ kΩ, $C = 0.02$ μF?

13. What is the discharge time of the circuit of problem 12?

14. What is the frequency value of the circuit of problem 12?

SUGGESTED LABORATORY ACTIVITIES

1. *Open-loop op-amp circuit.*

    a) Construct an open-loop op-amp circuit like the one shown in Figure 13–5. Use the following component values: $+V_{CC} = 10$ V, $-V_{CC} = 10$ V, $R_L = 2$ kΩ.

    b) Apply a signal from an AC signal source.

    c) With a meter or oscilloscope, measure input voltage (V_{in}) and output voltage (V_{out}).

    d) Calculate the voltage amplification (V_A).

2. *Noninverting op-amp circuit.*

    a) Construct the noninverting op-amp circuit shown in Figure 13–8.

b) Apply a signal from an AC signal source.

c) With an oscilloscope, measure V_{in} and V_{out} with the DC signal source adjusted to 0.

d) Calculate the AC voltage gain.

e) Disconnect the AC signal source and adjust the DC signal source until maximum V_{out} is obtained.

f) Calculate the DC voltage gain.

3. *Inverting op-amp circuit.*

a) Construct an inverting op-amp circuit similar to the one shown in Figure 13–9.

b) Perform the same procedure outlined in Lab 2 to calculate AC and DC voltage gain.

4. *IC power amplifier circuit.*

a) Construct the IC power amplifier circuit shown in Figure 13–11.

b) Use R_1 to adjust the AC input and test the circuit for operation.

5. *IC astable multivibrator circuit.*

a) Construct the IC astable multivibrator circuit shown in Figure 13–13. Use $+V_{CC} = 5$ V.

b) Connect an oscilloscope across the output and ground to observe the output.

c) Calculate the charge time, discharge time, and frequency.

d) Make a sketch of the output waveform.

e) Compare the measured waveform values to your calculations.

6. *IC 4-bit binary counter circuit.*

a) Construct the IC 4-bit binary counter shown in Figure 13–23.

b) Apply a square-wave pulse input to pin 14.

c) Observe the counting action of the circuit on the four LEDs.

7. *IC BCD counter circuit.*

a) Construct the IC BCD counter circuit shown in Figure 13–18.

b) Apply a square-wave pulse input to pin 14.

c) Observe the counting action of the circuit on the four LEDs.

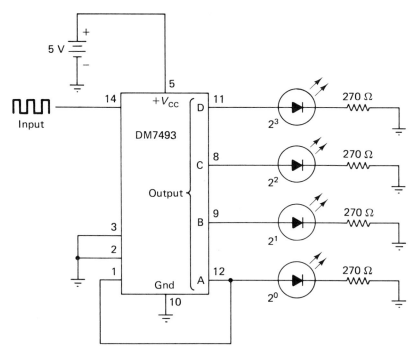

FIGURE 13–23

8. *Seven-segment digital display circuit.*

   a) Construct the seven-segment digital display circuit shown in Figure 13–19.

   b) Alternately connect each of the 470-Ω resistors to ground.

   c) Label the pin number that causes illumination of each display segment.

   d) Determine the pin combination which must be grounded to produce the numbers 0–9. Test each combination for operation.

9. *IC counting-timing circuit.*

   a) Construct the IC counting-timer circuit shown in Figure 13–22. Choose a value of capacitor (C) to provide the desired pulse rate.

   b) Close the switch to apply the clock pulse to pin 14 of the SN 7490 BCD counter.

   c) Observe the action of the MAN-1 seven-segment display.

CHAPTER 14
Introduction to Microcomputers and Microprocessors

This chapter provides a basic introduction to *microcomputers and microprocessors*, recent innovations in computer electronics technology. Computer technology has brought about significant changes in electronics. *Computers* have been used for a number of years to perform many operations. Calculations can be performed quickly, data can be stored and retrieved rapidly, and decisions can be made from data. All of us have become more dependent upon the computer.

In the mid-1960s *minicomputers* were developed to provide computer technology for applications not requiring the capacity of large-scale computers. These units were significantly less costly than large computers. Instant success demonstrated the usefulness of this computer with a wide range of applications. *Cathode-ray tube (CRT) displays* were used with minicomputers to present an immediate information display for the user. Information is ordinarily placed into this type of unit through a typewriter keyboard.

MICROCOMPUTER BASICS

Microcomputers were introduced by several manufacturers in the early 1970s. These computers were an extension of the technology employed by the minicomputer. Large-scale integration (LSI) technology was used to place thousands of discrete solid-state components on a single integrated circuit (IC) chip. The entire central processing unit (CPU) of a microcomputer is achieved today by a single IC chip called a *micropro-cessor unit (MPU)*. A microprocessor, with memory, input-output inter-face chips, and a power source, forms a microcomputer system. This combination of components can be built on a single printed circuit board and easily housed in a small cabinet. Microcomputer technology has revolutionized electronics by providing inexpensive small-capacity computers that can be used for many applications. Some examples of microcomputer applications are shown in Figure 14–1.

COMPUTER BASICS

The term *computer* is a general term that is used to cover a number of functions. It primarily refers to a system that will perform automatic computations when provided the appropriate information. Computers

FIGURE 14–1   (A) Microprocessor-based measuring instrument (Courtesy of Dranetz Technologies, Inc.).

range from pocket calculators to complex centralized units that are used to serve an entire industry. Electric energy is needed to energize the unit and operation is performed through electronic components. Information is provided in two states—*digital* or *analog*.

Digital computers use numbers represented by the presence or absence of voltage. These numbers were discussed in Chapter 12. A high voltage level or pulse usually represents a 1 state while no voltage indicates a 0 state. A voltage value or single pulse is described as a *bit* of information. The term bit is a contraction of the letters *bi* from the word *binary* and *t* from *digit*. A binary number has two states, or conditions, of operation. A group of pulses or voltage level changes produces a *word*. A *byte* consists of 8 bits.

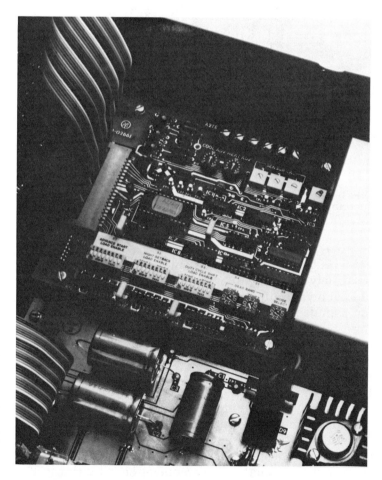

FIGURE 14–1 *(cont.)* (B) Microprocessor-based energy management system (Courtesy of Pacific Technology, Inc.).

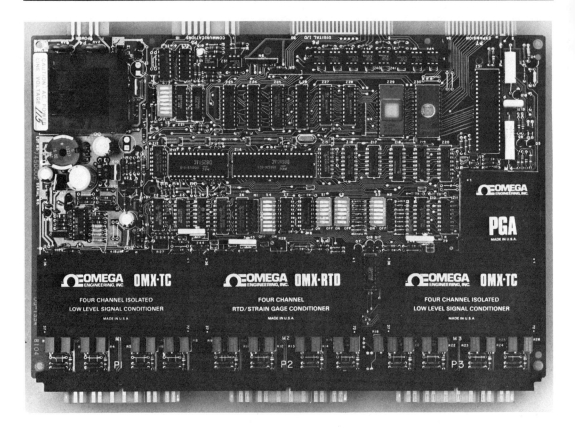

FIGURE 14–1 (cont.) (C) Microprocessor-based signal conditioner (Courtesy of Omega Engineering, Inc., Stamford, CONN 06907, U.S.A.).

Computer systems have certain fundamental parts. These parts may be arranged in a variety of different ways and still achieve the same function. The internal organization and design of each circuit differ considerably between manufacturers. Figure 14–2 shows a block diagram of a digital computer.

A digital computer is an electronic unit that consists of input and output devices, arithmetic logic and control circuitry, and some form of memory. Before the computer is placed into operation, a special set of instructions called a *program* must be supplied. These instructions are written in a *language* that the computer understands. This information is then supplied to the input unit by punched cards, punched tape, magnetic tape, or typewriter keyboard.

Input data supplied to the computer is translated into some type of number code, or *machine language*, before operation progresses. The arithmetic logic control unit then manipulates the input data according

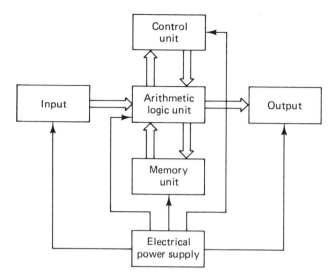

FIGURE 14–2   Digital computer system.

to its programmed instructions. After all the internal operations are complete, the coding process is reversed. The machine language is then translated back into another language. This information is used to actuate the output device.

The *memory* of a computer serves as a place to store the operating instructions that direct the CPU. Coded data in the form of 1s and 0s are "written" into memory through directions provided by a program. The CPU then "reads" these instructions from memory in a logical sequence and uses them to initiate processing operations. When the program is logical, processing proceeds in the correct manner with useful results.

Computers in operation today have one or more *output ports* that permit the CPU to communicate with the operator. Video display terminals (VDTs) for monitoring, line printers, and magnetic tape storage are typical output devices. Operational speed is an important characteristic of the output device.

The parts of a computer are basically the same for large-scale computers, minicomputers, and microcomputers. The differences between these units are in physical size. *Large-scale computers* usually employ thousands of discrete components and logic gates. This type of unit usually has an enormous memory capacity, high-speed operation, and simplified program routines. Typically, 32 or more bits of information are used to make programming words.

Minicomputer systems are made from logic gates that are arranged on numerous printed circuit boards. This type of unit has reduced memory capacity, requires a longer cycle time, and is less expensive. They are primarily designed for applications not requiring the capacity of a large-scale computer. Program word size is normally 8–16 bits.

Microcomputer systems are small units built around a single IC chip. The *microprocessor* of this system is a single-chip IC package that performs the arithmetic logic and control functions. These units are relatively inexpensive and can be used for many applications. They are somewhat more difficult to program than larger computers because of the reduced word size of 4–16 bits. They also have reduced memory capacity. Microcomputer systems are generally used to serve a specific need, solve a specific type of problem, or handle one application. This type of computer is often classified as a *dedicated* system since it is designed for a specific application.

MICROCOMPUTER SYSTEMS

Microcomputer systems are very significant developments in the field of electronics. The capabilities of the microcomputer have not been fully realized at this time. A microcomputer is complex when viewed in its entirety. It has a microprocessor, memory, an interface adapter, and several distribution paths called *buses*. To simplify this system, it is better to first look at a block diagram of its physical makeup, as shown in Figure 14–3.

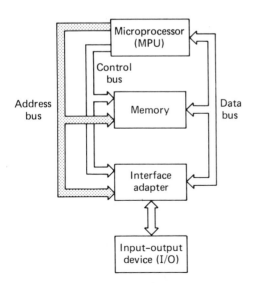

FIGURE 14–3   Simplified block diagram of microcomputer.

This simplified diagram of the microcomputer is similar to the basic computer diagram of Figure 14–2. The microprocessor has replaced the arithmetic logic unit and control unit in the microcomputer diagram. Information within this system is of two types: instructions and data. In a simple addition problem such as $9 + 2 = 11$, the numbers 9, 2, and 11 are data and the plus sign is an instruction. The data is distributed by the *data bus* to all parts of the system. Instructions are distributed by the *control bus* through a separate path. The *address bus* of the unit forms an alternate distribution path for the distribution of address data. It is normally used to place information into memory at the correct address. By an appropriate command from the microprocessor, data may be removed from memory and distributed to the output.

THE MICROPROCESSOR

A microprocessor (MPU) is the arithmetic logic unit and control section of a microcomputer which fits into a single IC chip. Typical chip sizes are approximately $\frac{1}{2}$ cm^2 and contain thousands of transistors, resistors, and diodes. Many companies are now manufacturing these chips in different designs and prices. Applications of this device are increasing at a remarkable rate.

A microprocessor is a digital device that is designed to receive data in the form of 1s and 0s. It may then store this data for future processing, perform arithmetic and logic operations in accordance with stored instructions, and deliver the results to an output device. In a sense, a microprocessor is a computer on a chip. The block diagram of a typical microprocessor, shown in Figure 14–4, contains a number of components connected in a rather unusual manner. Included in its construction are the arithmetic logic unit (ALU), an accumulator, a data register, address registers, program counter, instruction decoder, and a sequence controller.

ARITHMETIC LOGIC UNIT

All microprocessors contain an arithmetic logic unit (ALU). The ALU is a calculator chip that performs mathematical and logic operations on the data words supplied to it. It is made to work automatically by control signals developed in the instruction decoder. The ALU combines the contents of its two inputs, which are called the *data register* and the *accumulator*. Addition, subtraction, and logic comparisons are the primary operations performed by the ALU. The specific operation to

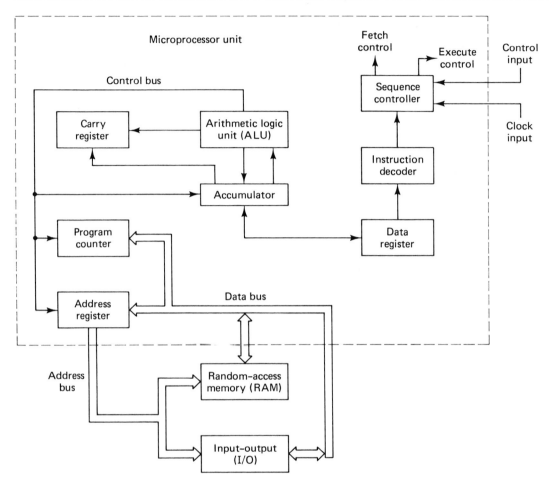

FIGURE 14–4   Simplified microprocessor block diagram.

be performed is determined by the control signal supplied by the *instruction decoder.*

The data supplied to the inputs of an ALU is normally in the form of 8-bit binary numbers. Upon receiving this data at the input, it is combined by the ALU in accordance with the logic of binary arithmetic. Since a mathematical operation is performed on the two data inputs, they are often called *operands.*

To demonstrate the operation of the ALU, assume now that two binary numbers are to be added—6_{10} and 8_{10}. Initially, the binary number 00000110_2 (6_{10}) is placed in the accumulator. The second operand, 00001000_2 (8_{10}), is then placed into the data register. When a proper control line to the ALU is activated, binary addition is performed, producing an output 00001110_2, or 14_{10}, which is the sum of the two operands. This value is then stored in the accumulator. It replaces the operand that appeared there originally. The ALU only responds to binary numbers.

ACCUMULATORS

The accumulators of a microprocessor are temporary registers designed to store operands that are to be processed by the ALU. Before the ALU can perform, it must first receive data from an accumulator. After the data register input and accumulator input are combined, the logical answer or output of the ALU appears in the accumulator.

A typical *instruction* for a microprocessor is LOAD ACCUMULATOR. This instruction enables the contents of a particular memory location to be placed into the accumulator. A similar instruction is STORE ACCUMULATOR. This instruction causes the contents of the accumulator to be placed in a selected memory location. The accumulator serves as an input source for the ALU and as a destination for its output.

DATA REGISTERS

The data register of a microprocessor serves as a temporary storage location for information applied to the *data bus*. Typically this register accommodates an 8-bit data word. An example of a function of this register is operand storage for the ALU input. It may also be called upon to hold an instruction while the instruction is being decoded. Another function is to temporarily hold data prior to the data being placed in memory.

ADDRESS REGISTERS

Address registers are used in microprocessors to temporarily store the address of a memory location that is to be accessed for data. In some units this register may be programmable. This means that it permits instructions to alter its contents. The program can also be used to build an address in the register prior to executing a memory reference instruction.

PROGRAM COUNTER

The program counter of a microprocessor is a memory device that holds the address of the next instruction to be executed in a program. This unit counts the instructions of a program in sequential order. When the MPU has fetched instructions addressed by the program counter, the count advances to the next location. At any given point during the sequence the counter indicates the location in memory from which the next information will be derived.

The numbering sequence of the program counter may be modified so that the next count may not follow a numerical order. Through this procedure, the counter may be programmed to jump from one point to another in a routine. This permits the MPU to have branch programming capabilities.

INSTRUCTION DECODERS

Each specific operation that the MPU can perform is identified by binary numbers known as *instruction codes*. Eight-bit words are commonly used for this code. There are 2^8 (256_{10}) separate or alternative operations that can be represented by this code. After a typical instruction code is pulled from memory and placed in the data register, it must be decoded. The instruction decoder examines the coded word and decides which operation is to be performed by the ALU. The output of the decoder is first applied to the sequence controller.

SEQUENCE CONTROLLERS

The sequence controller performs several functions in the operation of a microprocessor. Using clock inputs, it maintains the proper sequence of events required to perform a processing task. After instructions are received and decoded, the sequence controller issues a control sign that initiates the proper processing action. In most units the controller has the capability of responding to external control signals.

BUSES

The registers and other functional circuits of most microprocessors are connected together by a common bus network. The term *bus* is defined as a group of conductor paths that are used to connect data words to various registers. A diagram of registers connected by a common bus line is shown in Figure 14–5.

An advantage of bus-connected components is the ease with which a data word can be transferred or loaded into registers. Each register has inputs labeled clock, enable, load, and clear. When the load and enable input lines are at 0, each register is isolated from the common bus line.

To transfer a word from one register to another, it is necessary to make the appropriate inputs convert to the 1 state. To transfer the data of register A to register D, enable A (EA) and load D (LD) inputs both are placed in the 1 state. This causes the data of register A to appear on the common bus line. When a clock pulse arrives at the common inputs, the transfer is completed.

The word length of a bus is based upon the number of conductor paths. Buses for 4, 8, and 16 bits are commonly used in microprocessors. Many MPUs use an 8-bit data bus and a 16-bit address bus.

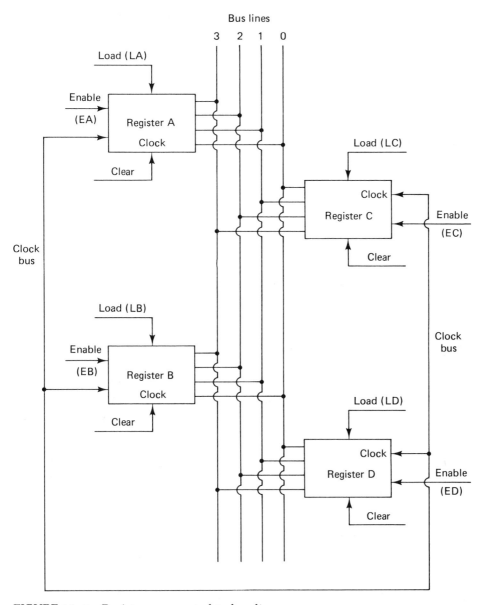

FIGURE 14–5   Registers connected to bus lines.

MICROPROCESSOR ARCHITECTURE

The physical layout, or "architecture," of a microprocessor is rather complex.

The MC6800 is one type of microprocessor. It is housed in a 40-pin dual-in-line package shown in Figure 14–6. Two accumulators, an index register, a program counter register, a stack pointer register, and a condition code register are included in its physical makeup.

MC6800

ACCUMULATORS

The MC6800 MPU has two accumulators labeled AA and AB. Each accumulator is 8 bits wide. Accumulators are designed to hold operands or data from the ALU.

Program instructions for each accumulator includes a mnemonic of its name and the operation to be performed. The loading of accumulator A is LDAA, and LDAB is the mnemonic for loading accumulator B. Storing data in accumulator A is STAA, with STAB achieving the same operation for accumulator B. With two accumulators in the MPU, arithmetic and logic operations can be performed on two different numbers at the same time without shifting to memory.

INDEX REGISTER

The index register of the MC6800 has a 16-bit, or 2-byte, capacity that is used to store memory addresses. This register has the capability of being loaded from two adjacent memory bytes. This allows its contents to be stored in two adjacent memory locations. Through this feature, data can be moved in 2-byte groups. The contents value increases by 1 when given the increment index register instruction or INX. A DEX instruction applied to the unit causes it to decrease by 1. This latter instruction is called *decrementing* the index register.

PROGRAM COUNTER

The program counter (PC) of the MC6800 is a 16-bit register that holds the address of the next byte to be fetched from memory. It can accommodate 2^{16}, or $65,536_{10}$, different memory addresses. Two 8-bit bytes are used for obtaining a specific address location.

STACK POINTER

The stack pointer (SP) of the MC6800 is a special 16-bit (2-byte) register. It uses a section of memory that has a last-in, first-out capability. With this capability the status of the MPU registers may be stored when branch or interrupt program subroutines are being performed.

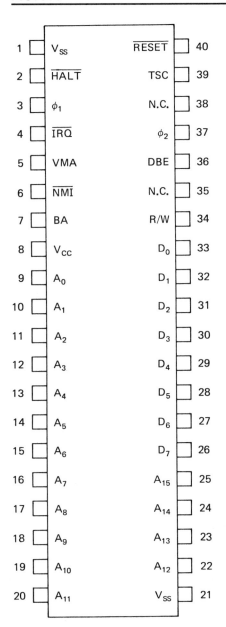

FIGURE 14–6   Pin layout of MC6800 microprocessor chip.

An address in the SP is the starting point of sequential memory locations in memory where the status of MPU registers is stored. After the
register status has been placed into the SP, it is decremented. When the
SP is accessed, the status of the last byte placed on the stack serves as
the first byte to be restored.

MEMORY

Microcomputer applications range from single-chip microprocessor units to complex systems that employ several chips. The primary difference in this range of applications is in the *memory* capabilities of the system. Single-chip microprocessors are limited in the amount of memory they possess. Additional memory is achieved economically through the use of auxiliary chips. The capabilities of a microcomputer system are extended by the range of memory they have.

Memory refers to the capability of a device to store logical data in such a way that a single bit or a group of bits can be easily accessed or retrieved. Memory is achieved in a variety of different ways. Microcomputer systems usually have read-write memory and read-only memory. These two types of memory are ordinarily accomplished by semiconductor circuits on a single IC chip.

READ-WRITE MEMORY

Read-write memory is widely used in microcomputer systems today. Read-write chips of the large-scale integration (LSI) type are capable of storing 16,384, or 16K, bits of data in an area less than one-half of a square centimeter. The structure of this chip includes a number of discrete circuits, each having the ability to store binary data in an organized rectangular configuration. Access to each memory location is provided by coded information from the microprocessor address bus. The read-write function indicates that data can be placed into memory or retrieved at the same rate.

A simplified representation of the memory process is illustrated by the 8×8 memory unit of Figure 14–7. Memory ICs are usually organized in a rectangular pattern of rows and columns. Figure 14–7 shows eight rows that can store 8-bit words, or 64 single bits of memory. To select a particular memory address, a 3-bit binary number is used to designate a specific row location, and three additional bits are used to indicate the column location. In this example the row address is 3_{10}, or 011_2, and the column address is 5_{10}, or 101_2. The selected memory cell address is at location 30.

Eight-bit word storage is achieved in the memory of Figure 14–7 by energizing one row and all eight columns simultaneously. The row and column decoders perform this operation. Some of the read-write memory units available today have capacities of 8×8, 32×8, 128×8, 1024×8, and 4096×8.

To write a word into memory, a specific address is first selected according to the data supplied by the address bus. Reading data from memory does not destroy the data at each cell. Data reading can take place as long as the memory unit is energized electrically. A loss of elec-

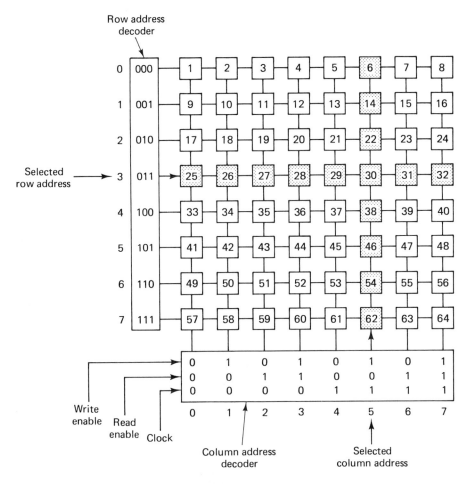

FIGURE 14–7   8 × 8 simplified memory unit.

trical power, or turning the unit off, *destroys* the data placed at each memory location. Solid-state read-write memory is classified as *volatile* because of this characteristic.

READ-ONLY MEMORY

Most microcomputer systems require memories that contain permanently stored or rarely altered data. Examples of this are math tables and permanent program data. Storage of this type is provided by *read-only memory (ROM)*. Information is often placed in this type of memory unit when the chip is manufactured. ROM data is *nonvolatile*, which means that it is not lost when the power source is turned off.

A variation of the ROM is the *read-mostly memory*. This unit is used where read operations are needed more frequently than write operations. *Programmable read-only memory (PROM)* is another type that can be optically erased by exposure to an ultraviolet light source. After light exposure each memory cell of the unit goes to a 0 state. Writing data into the chip again initiates a new program.

An alternative to the optically erasable read-mostly memory is the *electrically altered read-only memory* (EAROM) or the erasable programmable read-only memory (EPROM). This type of chip permits erasures of individual cells or word locations instead of the entire chip. Memory cell structure is very similar to the optically erasable ROM. The potential usefulness of the EAROM and EPROM is considerable for several applications.

MICROCOMPUTER FUNCTIONS

There are certain functions that are basic to most microcomputer systems. Among these functions are timing, fetch and execution, read memory, write memory, input-output transfer, and interrupts. A knowledge of these basic functions is important for understanding the operation of microcomputers.

TIMING

The operation of a microcomputer system is achieved by a sequence of cycling instructions. The MPU fetches an instruction, executes the required operations, fetches the next instruction, executes it, and continues to operate in a cycling pattern. This means that all actions occur at or during a precisely defined time interval. An orderly sequence of operations like this requires some type of precision clock mechanism. A free-running electronic oscillator or clock is responsible for this function. In some systems the clock may be built into the MPU, while in others timing units feed the system through a separate clock control bus. The sequence of operations is controlled by a timing signal.

Sequential operations such as fetch and execute are achieved within a period called the MPU cycle. The fetch portion of this cycle consists of the same series of instructions. It therefore takes the same amount of time for each instruction. The execute phase of an operation may consist of many events and sequences, depending upon the specific instruction being performed. This portion of the cycle varies a great deal with the instruction being performed.

The total interval taken for a timing pulse to pass through a complete cycle from beginning to end is called a *period*. One or more clock periods may be needed to complete an operational instruction.

FETCH AND EXECUTE OPERATIONS

After programmed information is placed into the memory of a microcomputer, its action is directed by a series of fetch and execute operations. This sequence of operations is repeated until the entire program has cycled to its conclusion. This program tells the MPU specifically what operations it must perform.

The operation begins when the start function is initiated. This signal actuates the control section of the MPU, which automatically starts the machine cycle. The first instruction that the MPU receives is to fetch the next instruction from memory. The MPU may then issue a read operational code instruction. The contents of the program counter are then sent to the program memory, which returns the next instruction word. The first word of this instruction is then placed into the instruction register. If more than one word is included in an instruction, a longer cycling time is needed. After the complete instruction word is in the MPU, the program counter is incremented by one count. The instruction is then decoded, which prepares the unit for the next fetch instruction.

The execution phase of operation is based upon which instruction is to be performed by the MPU. This instruction may be read memory, write in memory, read the input signal, transfer to output, or an MPU operation such as add registers, subtract, or register-to-register transfer. The time of an operation is dependent upon the programmed information placed in memory.

The popular 8080 microprocessor takes a number of the clock periods to perform its operations. This particular chip can be operated at a clock rate of 2 MHz. A single cycle of the clock has a period 1/2,000,000 or 0.0000005 s. Periods this small are best expressed in microseconds (0.5 μs) or nanoseconds (500 ns). The fastest instruction change that can be achieved by this chip requires 4 clock periods, or 4×500 ns = 2 μs. Its slowest instruction requires 18 periods, or 18×500 ns = 9 μs. The operational time of an MPU is a good measure of its effectiveness and how powerful it is as a functional device.

READ MEMORY OPERATION

The read memory operation is an instruction that calls for data to be read from a specific memory location and applied to the MPU. To initiate this operation, the MPU issues a read operation code and sends it to the proper memory address. The read-write memory unit then sends the data stored at the selected address into the data bus. This 8-bit number is fed to the MPU, where it is placed in the accumulator after timing pulse has been initiated.

WRITE MEMORY OPERATION

The write memory operation is an instruction that calls for data to be placed into or stored at a specific memory location. This function is initiated when the MPU issues a write operation code and sends it to a selected read-write memory unit. Data from the data bus is then placed into the selected memory location for storage. In practice, 8-bit numbers are usually stored as words in the memory unit.

INPUT-OUTPUT TRANSFER OPERATION

In a microcomputer system, *input-output (I/O) transfer* operations are very similar to read-write operations. The major difference is the opcode data number used to call up the operation. When the MPU issues an input-output opcode, it actuates the appropriate *I/O port* which either receives data from the input or sends it to the output device according to the coded instruction.

A simplified microcomputer system with a read-write memory, read-only memory, output, and input connected to a common address bus and data bus is shown in Figure 14–8. In this type of system data is processed by the MPU. Data moves in either direction to the read-write memory and flows from the ROM or input into the data bus. The output flows from the data bus to the output device.

INTERRUPT OPERATION

Interrupt operations are often used to improve the efficiency of a microcomputer system. Interrupt signals are generated by peripheral equipment such as keyboards, displays, printers, or process control devices. These signals are generated by peripheral equipment and applied to the MPU. They inform the MPU that a peripheral device needs some type of attention.

Assume that a microcomputer system is designed to process a large volume of data and the output is a *line printer*. The MPU can output data at a very high rate of speed compared to the time needed to actuate a character on the line printer. This means that the MPU has to remain idle while waiting for the printer to complete its operation.

If an interrupt is used by the MPU of a microcomputer system, it can output a data byte and then return to processing data while the printer completes its operation. When the printer is ready for the next data byte, it requests an interrupt. Upon acknowledging the interrupt, the MPU stops program execution and automatically branches to a subroutine that will output the next data byte to the printer. The MPU then continues the main program execution where it was interrupted.

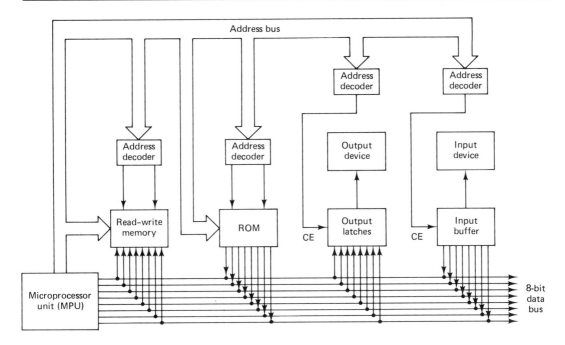

FIGURE 14–8   Simplified MPU address bus and data bus.

Through this procedure high-speed MPU operation is not restricted by slow-speed peripheral equipment.

PROGRAMMING

The term *program* refers to a series of instructions developed for a computer that permit it to perform a prescribed operation. In microcomputer systems some programs are hard-wired, some are in read-only memories called firmware, and others are described as software. A *dedicated system (hard-wired)* designed for a specific function is usually hardware programmed so that it cannot be adapted for other tasks without a circuit change. *Firmware systems* have programmed material placed on ROM chips. Program changes for this type of system can be achieved by changing the ROMs. A *software* type of program has the greatest flexibility. This type of program is created on paper and transferred to the system by a keyboard, punched cards, or magnetic tape. Instructions are stored at read-write memory locations and performed by calling for these instructions from memory.

The design of a microcomputer system and its purpose usually dictate the type of programming method used. Combined firmware and software programmed systems with a keyboard are very common.

INSTRUCTIONS

The instructions for a microcomputer system normally appear as a set of characters or symbols which are used to define a specific operation. These symbols are similar to those that appear on a typewriter. Included are decimal digits 0–9, the letters A–Z, and in some cases punctuation marks and specialized keyboard characters. Symbolized instructions may appear as binary numbers, hexadecimal numbers, or as mnemonic codes.

Microcomputers may employ any one of several different codes in a program according to its design. Each type of MPU has an *instructional set* it is designed to understand and obey. These instructions appear as binary data symbols and words. *Machine instructions* of this type are normally held in a read-only memory unit that is address selected and connected to the MPU through the common data bus. Instructional sets are an example of firmware because they are fixed at a specific memory location and cannot be changed.

Microcomputer instructions usually consist of 1, 2, or 3 bytes of data. This type of data must follow the instruction commands in successive memory locations. These instructions are usually called *addressing modes*.

One-byte instructions are often called *inherent-mode instructions*. These instructions are designed to manipulate data to the accumulator registers of the ALU. No address code is needed for this type of instruction because it is an implied machine instruction. The instruction CLA, for example, is a 1-byte *opcode* that clears the contents of accumulator A. No specific definition of data is needed, nor is an address needed for further data manipulation. This instruction simply clears the accumulator register of its data.

Inherent mode instructions differ a great deal between manufacturers. Some representative opcode instructions are shown in Figure 14–9. Note that the meaning of the instruction, its code in hexadecimal form, and a mnemonic are given for each instruction. Each microcomputer system has a number of instructions of the 1-byte type which contain only an opcode.

IMMEDIATE ADDRESSING

Immediate addressing is accomplished by a 2-byte instruction that contains an opcode and an operand. In this addressing mode the opcode

Mnemonic	Opcode	Meaning
ABA	1B	Add the contents of accumulators A and B. The result is stored in accumulator A. The contents of B are not altered.
CLA	4F	Clear accumulator A to all zeros.
CLB	5F	Clear accumulator B.
CBA	11	Compare accumulators: Subtract the contents of ACCB from ACCA. The ALU is involved but the contents of the accumulators are not altered. The comparison is reflected in the condition register.
COMA	43	Find the ones complement of the data in accumulator A, and replace its contents with its ones complement. (The ones complement is simple inversion of all bits.)
COMB	53	Replace the contents of ACCB with its ones complement.
DAA	19	Adjust the two hexadecimal digits in accumulator A to valid BCD digits. Set the carry bit in the condition register when appropriate. The correction is accomplished by adding 06, 60, or 66 to the contents of ACCA.
DECA	4A	Decrement accumulator A. Subtract 1 from the contents of accumulator A. Store result in ACCA.
DECB	5A	Decrement accumulator B. Store result in accumulator B.
LSRA	44	Logic shift right, accumulator A or B. $0 \rightarrow [b_7\ b_6\ b_5\ b_4\ b_3\ b_2\ b_1\ b_0] \rightarrow [C]$
SBA	10	Subtract the contents of accumulator B from the contents of accumulator A. Store results in accumulator A.
TAB	16	Transfer the contents of ACCA to accumulator B. The contents of register A are unchanged.
TBA	17	Transfer the contents of ACCB to accumulator A. The contents of ACCA are unchanged.
NEGA	40	Replace the contents of ACCA with its twos complement. This operation generates a negative number.
NEGB	50	Replace the contents of ACCB with its twos complement. This operation generates a negative number.
INCA	4C	Increment accumulator A. Add 1 to the contents of ACCA and store in ACCA.
INCB	5C	Increment accumulator B. Store results in AACB.
ROLA	49	Rotate left, accumulator A.

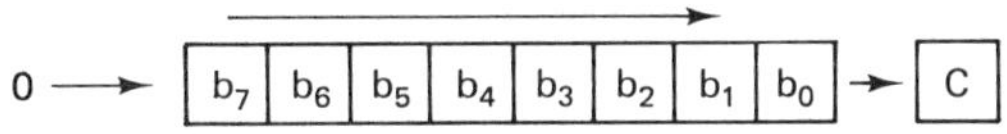

FIGURE 14–9 Opcode instructions.

appears in the first 8-bit byte, followed by an 8-bit operand. A common practice is to place intermediate addressing instructions in the first 256 memory locations. These instructions can be retrieved very quickly since this is the fastest mode of operation.

RELATIVE ADDRESSING

Relative addressing instructions are designed to transfer program control to a location other than the next consecutive memory address. In this type of addressing, two 8-bit bytes are used for the instruction. Transfer operations of this type are often limited to a specific number of memory locations in front or in back of its present location. The 2-byte instruction contains an opcode in the first byte and an 8-bit memory location in the second byte. The second byte points to the location of the next instruction which is to be executed. This type of instruction is used for branching programming operations.

INDEXED ADDRESSING

Indexed addressing is achieved by a 2-byte instruction and is very similar to the relative addressing mode of instruction. In this type of addressing, the second byte of the instruction is added to the contents of the index register to form a new, or "effective," address. This address is obtained during program execution rather than being held at a predetermined location. A newly created effective address is held in a temporary memory address register so that it will not be altered or destroyed during processing.

DIRECT ADDRESSING

Direct addressing is a common mode of instruction. In this type of instruction the address is located in the next byte of memory following the opcode. This permits the first 256 bytes of memory, from 0000 to $00FF_{16}$, to be addressed.

EXTENDED ADDRESSING

Extended addressing is a method of increasing the capability of direct addressing so that it can accommodate more data. This mode of addressing is used for memory locations above $00FF_{16}$ and requires 3 bytes of data for the instruction. The first byte is a standard 8-bit opcode. The second byte is an address location for the most significant or highest order of the data in 8 bits. This is followed by the third byte, which holds the address of the least significant or lowest 8 bits of the data number being processed.

PROGRAM PLANNING

A microcomputer system could not solve even the simplest type of problem without the help of a well-defined program. After the program has been developed, the system simply follows this procedure to accomplish the task. Programming is an essential part of nearly all computer system applications.

Before a program can be effectively prepared for a microcomputer system, the programmer must be fully aware of the specific instructions that are performed by the system. Microcomputer systems have a list of instructions that are used to control their operation. The "instruction set" of a microcomputer is the basis of all programs.

A programmer should be familiar with the instruction set of the system being used on a microcomputer. He should be able to decide what specific instructions are needed to solve a problem. A limited number of operations can usually be developed without the aid of a diagrammed plan of procedure. Complex problems require a specific plan in order to reduce confusion or to avoid the loss of an important operational step. Flowcharts are commonly used to aid the programmer in this type of planning. Figure 14–10 shows some of the flowchart symbols that are used in program planning.

When programming a microcomputer system, the programmer must be aware of specific instructions. He must decide upon what instructions are needed to solve a problem, plan a flowchart, develop a programming sheet, initiate the program, correct it if needed, and exe-

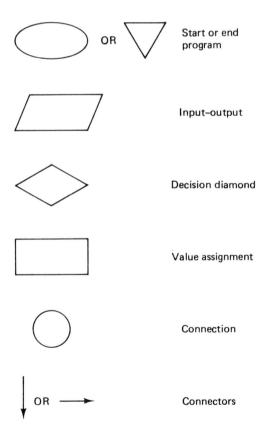

FIGURE 14–10   Programming flowchart symbols.

cute its operation. Microcomputers may be used for practically unlimited applications.

In this chapter an introduction to microcomputers and microprocessors has been presented. Answer the following review questions to better understand the material in this chapter.

REVIEW

1. Compare the functional differences among a central computer, a minicomputer, and a microcomputer system.

2. Explain what is meant by the terms *bit*, *byte*, and *word*.

3. What are the fundamental parts of a computer system?

4. What are the functions of the following microcomputer parts?

    a) Bus

    b) MPU

    c) ALU

    d) Accumulator

    e) Data register

    f) Address register

    g) Program counter

    h) Instruction decoder

    i) Sequence controller

5. Discuss the MC 6800 microprocessor unit.

6. Compare the read-write and ROM memories.

7. Why is a read-write memory considered to be volatile?

8. What is an EAROM? EPROM?

9. Why is timing an essential microcomputer function?

10. Explain the fetch and execute operations of a microcomputer.

11. How is data transferred in a microcomputer system between I/O ports?

12. What is the purpose of the interrupt operation of a microcomputer system?

13. What is an instructional set for a microcomputer?

14. Explain the differences among the following:

    a) Inherent-mode instructions

    b) Immediate addressing

    c) Relative addressing

    d) Indexed addressing

    e) Direct addressing

CHAPTER 15
Miscellaneous Devices and Circuit Applications

In this chapter, devices and circuits that have not been studied in previous chapters but are much used in the field of electronics will be examined. These devices include the cathode-ray tube, thermistors, and relays, among others. Appropriate electronic circuitry utilizing these devices is also examined.

CATHODE-RAY TUBE

The *cathode-ray tube (CRT)* is a vacuum tube that is used in all oscilloscopes to provide a visual display of electric energy. Basically the CRT is a large glass tube from which all oxygen has been removed. Within the glass tube are found many elements which control the action of a stream of electrons as these pass from one end of the tube to the other. Similar to most vacuum tubes, these electrons originate as a result of thermionic emission.

Figure 15–1 illustrates the basic elements of the CRT as well as the location of each element within the tube's enclosure. Like most vacuum tubes, the CRT is plugged into an appropriate socket which provides the means for the external electrical connections to be made to the tube's internal elements.

The CRT's *filament* is heated due to an external voltage causing current to flow through the filament. This heat from the filament is trans-

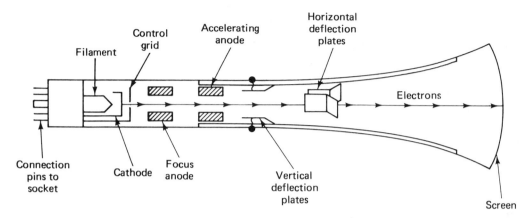

FIGURE 15–1   Cathode-ray tube construction characteristics.

ferred to the tube's cathode which emits electrons. These electrons are forced to move through and are controlled by the tube's control grid, focus anode, accelerating anode, vertical deflection plates, and horizontal deflection plates due to the very positive accelerating anode and Aquadag (graphite coating on the inside of the tube).

As the electrons from the cathode pass through the control grid, which is negatively charged, they are forced to pass through a small hole located near the center of the grid. This tends to form the electrons into a more compact beamlike structure.

As the electrons pass from the control grid through the focusing and accelerating anodes, they are formed into a very well defined beam. This is often referred to as *electron convergence* and is due primarily to the *electrostatic* difference of potential that exists between these elements. The electron beam then passes between both the vertical and horizontal deflection plates.

The vertical deflection plates are made positive and negative by virtue of external connection posts located on the tube's outer wall. As the top vertical deflection plate becomes more positive as compared to the bottom plate, the electron beam, which is negative, is attracted toward the top vertical plate and repelled by the bottom plate. Likewise, as the bottom vertical deflection plate becomes more positive than the top plate, the electron beam is repelled by the top plate and is attracted toward the bottom plate. Thus, by controlling the electrical potential of the vertical deflection plates, the electron beam is caused to move up and down on a vertical plane.

The horizontal deflection plates are made positive and negative by similar external connections. As the right horizontal deflection plate becomes more positive than the left, the electron beam is repelled from the left and attracted toward the right. Likewise as the left horizontal deflection plate becomes more positive than the right, the electron beam is repelled from the right and attracted to the left. Thus, the electron beam is moved from left to right or right to left on a horizontal plane.

By controlling the electrical potentials on both the vertical and horizontal deflection plates the electron beam is moved up and down as it is moved from left to right. As the beam strikes the phosphor coating on the screen (causing the affected phosphor to glow), it creates a pattern, or *trace*, identical to the external electrical potential found on the deflection plates. Some common traces are illustrated in Figure 15–2.

THERMISTORS

Thermistors are temperature-sensitive resistors that exhibit a *negative temperature coefficient*. The electrical resistance of the thermistor is reduced when it is placed in an environment of higher temperature. Likewise, the resistance of the thermistor is increased when its environmental temperature is decreased.

These temperature-sensitive devices are generally classified as semiconductors that are formed from oxides (iron, cobalt, copper) into beads, washers, rods, or discs with external terminals or leads suitable for electrical connections. Thermistors are classified as being "directly" or "indirectly" heated. Figure 15–3 illustrates the electrical symbols and physical characteristics of both types of thermistors.

A directly heated thermistor is one that is placed directly in the environment to which it is to respond. Its resistance is controlled by the temperature of its surroundings.

The indirectly heated thermistor is enclosed in an evacuated glass bulb along with a heater. The resistance of the indirectly heated thermistor is controlled by the temperature of the heater, which in turn is controlled by the current flowing through the heater due to the heater's external voltage.

There are three major characteristics of thermistors that make them useful as electrical devices: resistance-temperature, voltage-current, and current-time characteristics.

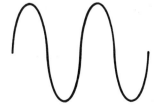

FIGURE 15–2   Common CRT patterns.

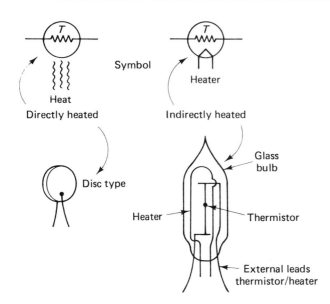

FIGURE 15–3   Characteristics of thermistors.

RESISTANCE-TEMPERATURE CHARACTERISTICS

The relation between a thermistor's resistance and the temperature of its environment is inversely proportional. That is, as the temperature of the thermistor increases, its resistance will decrease. This characteristic enables the thermistor to be used, along with other electronic components and circuitry, in temperature-sensing and temperature-measuring applications. Figure 15–4 illustrates a thermistor bridge circuit that is used to indicate the temperature of an environment.

In this example, the resistance of T_1 and T_2 are the same. R_1 is adjusted until V_1 (zero-centered voltmeter) indicates a *null* or zero reading. This indicates that the bridge circuit is balanced. The temperature, and thus the resistance, of T_1 is carefully controlled and remains stable. When T_2 is placed in an environment with a different temperature as compared to the temperature of the environment of T_1, the bridge becomes unbalanced. This is due to the change of T_2's resistance caused by the temperature of the new environment. The magnitude of the unbalanced state of the bridge circuit is indicated by the zero-centered voltmeter (V_1). If the meter is calibrated in degrees Fahrenheit or Celsius, the difference in temperature between T_1 and T_2 is directly measured. Since V_1 is zero-centered, a temperature difference above or below that of T_1 is indicated.

Figure 15–5 illustrates a common combination of components connected in such a way as to allow the measurement of temperatures above

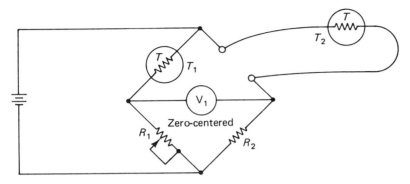

FIGURE 15–4   Thermistor bridge.

72 °F or 22.2 °C (room temperature) or any other chosen temperature that is within a thermistor's range (-60 to $+300$ °C).

In Figure 15–5, R_1 is adjusted until the collector current of Q_1 is just at zero. This causes Q_1 to be biased at its cutoff point and to function as a Class B amplifier. As a result, M_1 indicates zero collector current. If the temperature of T_1 increases, its resistance will decrease. This will allow additional current to flow through R_1 and will result in the base of Q_1 becoming more positive. This forward biases Q_1 and causes collector current proportional to the resistance change of T_1. If M_1 is calibrated in degrees Fahrenheit or Celsius, the temperature increase in the environment of T_1 is directly measured by M_1.

The thermistor, along with appropriate electronic circuitry, can be used to measure the rate of gas or liquid flow within a closed chamber. Circuits employed to measure flow are very similar to those previously discussed and employ either a bridge or amplifier or both.

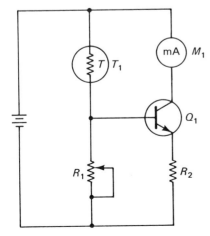

FIGURE 15–5   Temperature measurement circuit.

VOLTAGE-CURRENT CHARACTERISTICS

The voltage-current characteristic explains a thermistor's "self-heating" effect. The resistance of a thermistor depends upon its temperature. Recall from past discussions that the temperature of any conductor depends upon, among other considerations, the amount of current through that conductor. Normally, the greater the current flow through a conductor, the greater its temperature.

As expected, current flow through a thermistor results in a certain amount of power being converted to heat. This heat tends to lower the resistance of the thermistor and allows current to further increase. This causes more power to be converted to heat, resulting in an additional decrease in the thermistor's resistance. This effect is known as the thermistor's *self-heating effect* and is usually stated as its *dissipation constant*. The thermistor's dissipation constant is usually provided by the manufacturer and is an important consideration when extreme accuracy in temperature measurement is essential. Generally, the dissipation constant as recognized by most manufacturers of thermistors is the amount of electric power as stated in milliwatts that will increase the temperature of the thermistor 1 °C above its surroundings.

Normally, very low values of current flow through a thermistor will not cause self-heating. This results in the thermistor's voltage-current characteristic being somewhat linear. As the applied voltage is increased, resulting in an increased current through the thermistor, self-heating begins, causing a decrease in the thermistor's resistance. This decrease in the thermistor's resistance causes the thermistor's voltage to decrease as the thermistor's current increases. Figure 15–6 illustrates the thermistor's voltage-current characteristic.

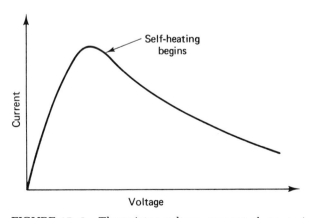

FIGURE 15–6   Thermistor voltage-current characteristics.

CURRENT-TIME CHARACTERISTICS

The current-time characteristic of the thermistor defines the response time required for the device to change its resistance when placed in a new environmental temperature, thus allowing a different level of current to flow. The response time of a thermistor is expressed as its *time constant*, the time required in seconds for it to change its temperature-resistance approximately 63 percent of the value of the temperature change of its new environment. Thus, if a thermistor is taken from an environmental temperature of 72 °F (22.2 °C) and placed in an environmental temperature of 172 °F (77.7 °C), the time required for the thermistor to change to exhibit a temperature of 135 °F (57.2 °C) is its time constant. This becomes clear because 172 °F (new temperature) − 72 °F (old temperature) = 100 °F, and 63 percent of 100 °F = 63 °F. Thus, 63 °F + 72 °F = 135 °F. Approximately five time constants are required for a thermistor to completely reflect a new environmental temperature as exhibited by its change in resistance and corresponding current.

RELAYS

Relays are classified as being either electromagnetic or solid state and are used as switches to control the operation of other devices or circuits in a time delay mode, sensing mode, or high-voltage mode.

Electromagnetic relays were used as the switching mechanisms in the first successful computers. Even though these devices have been replaced by solid-state electronic devices in computer applications, they are still used extensively in industrial and communication electronic applications. The solid-state relay was introduced in 1971. As implied by its name, there are no moving parts. Switching is achieved by employing semiconductor devices such as triacs, SCRs, transistors, and diodes.

ELECTROMAGNETIC RELAYS

Electromagnetic relays are devices that contain a coil, armature, and one or more sets of contacts. Figure 15–7 illustrates a very basic electromagnetic relay. There are two types of contacts normally associated with an electromagnetic relay. These are classified as normally open (NO) or normally closed (NC) contacts. When a sufficient voltage is applied to the relay's coil, a magnetic field across the coil is caused by current flow through the coil. This magnetic field attracts the armature and causes the NO contacts to close and the NC contacts to open. When the coil voltage is removed, the magnetic field across the coil collapses, the armature is no longer attracted to the coil, and the armature's

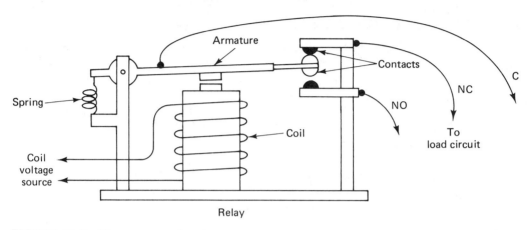

FIGURE 15–7   Electromagnetic relay construction.

spring causes the armature to return to its normal position. This allows the NC contacts to close and the NO contacts to open. Figure 15–8 illustrates several electrical symbols of the electromagnetic relay's coil and contacts.

Important characteristics concerning the electromagnetic relay's operation are the maximum current that can safely be switched by the relay's contacts, the relay coil's "make," or "pickup," current, and the relay coil's "break," or "dropout," current. The electromagnetic relay's *maximum contact current* is the largest amount of current that can be switched on or off by the relay's NO or NC contacts. This specification is usually supplied by the manufacturer and is determined by contact material, size, and placement. The relay coils *make, or pickup, current* is the minimum amount of current flow through the coil that will create a magnetic field that will pull the armature and cause the contacts to change states. The amount of coil current that will cause the coil's magnetic field to release the armature is known as the coil's *break, or drop-*

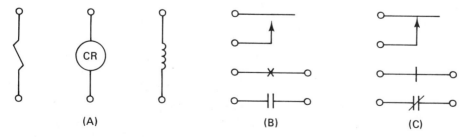

FIGURE 15–8   Relay electrical symbols. (A) Relay coils. (B) Normally open (NO) contacts. (C) Normally closed (NC) contacts.

out, *current*. Figure 15–9 illustrates a typical circuit employing a multi-contact (more than one) relay controlling the action of two loads.

In this circuit light causes PC1 to exhibit a low resistance. This causes a sufficient current to flow through R_1 and results in Q_1 being forward biased. The forward-biased state of Q_1 is sufficient to cause enough collector current to flow through the relay coil (CR) to allow relay pickup. Thus, the NO contacts close to energize load 2. The NC contacts open to de-energize load 1.

In the absence of light, PC1 resistance rises to a much higher level. This reduces the current through R_1 and causes Q_1 to be less forward biased. The collector current of Q_1 is reduced below the relay coil's dropout value, and the NO contacts open while the NC contacts close. This turns off load 2 and turns on load 1.

SOLID-STATE RELAYS (SSRs)

Solid-state relays (SSRs) are more reliable, faster, and require less maintenance (none) than the electromagnetic relays previously discussed. When used correctly, the SSR has an infinite operating life. Figure 15–10 illustrates a typical block diagram of an SSR. In comparing its operation to that of the electromagnetic relay, it can be seen that the control circuit performs the same function as the electromagnetic relay's coil. The solid-state switching device (SCR, transistor, triac) performs the same function as the electromagnetic relay's contacts.

Important specifications associated with the SSR are its *maximum load current* capability, *maximum load voltage* capability, *minimum*

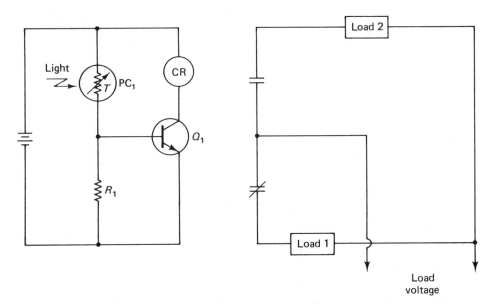

FIGURE 15–9 Typical relay control circuit.

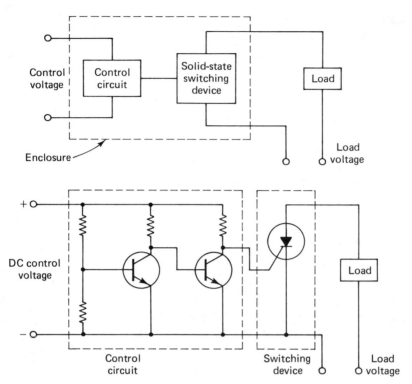

FIGURE 15–10   Solid-state relay.

control current required to turn on the control circuit, *minimum voltage* required to turn on the control circuit, and *turn on-off time*. Usually, the maximum control circuit current and voltage are also specified.

TRANSDUCERS

Although transducers are not solid-state devices, they often play a significant role in the operation of solid-state circuits. Devices that convert one form of energy into another form are referred to as *transducers*. Transducers are often used to convert physical quantities to electrical quantities. For example, a *thermocouple* is a transducer that converts heat energy into electric energy. Also, a microphone is a transducer that converts sound energy into electric energy. Numerous other examples of transducers are used in our homes as well as in industry. There are several basic types of transducers that may be used with electronic circuits. Common classifications of transducers include resistive, capacitive, inductive, piezoelectric, thermoelectric, and optoelectric.

RESISTIVE TRANSDUCERS

Resistive transducers convert variations of resistance into electrical variations. A common type of resistive transducer is the *strain gauge* shown in Figure 15–11A. Strain gauges are ordinarily used to sense a change in dimension of a material as the material is subjected to a stress. Most applications are confined to laboratory usage for testing material strength or as calibrating instruments. The strain gauge itself is constructed of fine gauge wire about 0.001 in. in diameter mounted into an insulating strip. The wire used has a high elasticity so that it will easily change dimension. When subjected to a stress, the wire will be stretched. Thus, the cross-sectional area of the wire is reduced and its length is increased. The resistance of a conductor can be expressed mathematically as

$$R = \rho \frac{l}{A}$$

where:

R is the resistance of the conductor,

ρ is the resistivity constant of the conductor,

l is the length of the conductor, and

A is the cross-sectional area of the conductor.

Therefore, as the wire of the strain gauge is stretched, its resistance will increase because of the change in cross-sectional area and length.

Another type of resistive transducer utilizes the potentiometer principle, as shown in Figure 15–11B and 15–11C. This type of transducer changes resistance when the position of its movable contact is changed. By increasing the length of wire between terminals A and B, the resistance between those two points is increased. This principle is often used to sense physical displacement by allowing the displacement to cause movement of the sliding contact.

Another application of resistive transducers is for electrochemical measurements. The resistivity of various chemical solutions changes due to a change of state that may be brought about by the quantity to be measured. For instance, pH (alkalinity or acidity) can be measured by passing an electric current through a solution containing that which is to be measured. The pH of the solution will cause the resistivity to vary, thus varying the current in the external circuit calibrated to measure pH.

CAPACITIVE TRANSDUCERS

Capacitive transducers rely on a change in capacitance brought about by the change in some physical quantity. Capacitance exists when

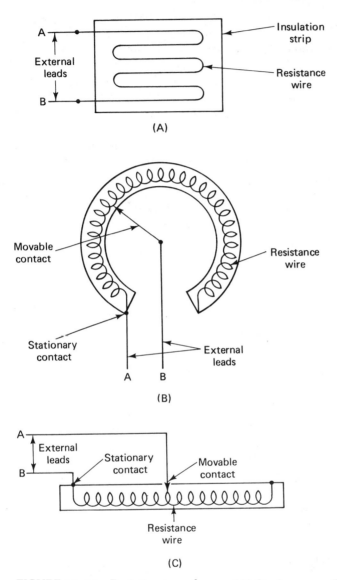

FIGURE 15–11   Resistive transducers. (A) Strain gauge. (B) Rotary potentiometric transducer. (C) Flat potentiometric transducer.

two conductive materials (plates) are separated by a dielectric material. Capacitance can be increased by increasing the area of the plates or by decreasing the thickness of the dielectric. One application of capacitive transducers illustrated in Figure 15–12 is for sensing fluid pressure. This type of transducer is placed into a fluid line. Plate 1 of the capacitor is a conductive diaphragm inserted into the fluid line to sense any varia-

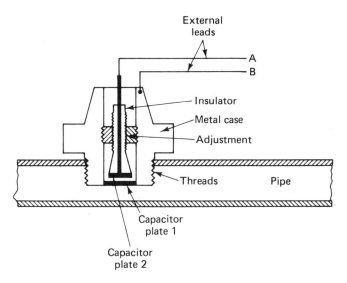

FIGURE 15–12   Capacitive transducers.

tion in fluid pressure. It is electrically connected to the housing. Plate 2 is mounted adjacent to plate 1 and is initially adjusted to calibrate the indicator scale. Plate 2 is held in position, while plate 1 will vary in position due to changes in pressure. When the pressure of the fluid in the line increases, plate 1 will move closer to plate 2. Since the distance between capacitor plates is decreased, the capacitance between terminals A and B will increase. Also, when pressure decreases, capacitance will decrease. Thus, variations in pressure cause changes in capacitance.

INDUCTIVE TRANSDUCERS

Inductive transducers can also be varied to cause an electrical change. Usually, inductive transducers such as the one shown in Figure 15–13A have a stationary coil and a movable core. The movable core can be connected to some physical variable whose movement is to be sensed. As the core changes position within the stationary coil, the inductance of the coil will vary. The current flow through the coil will vary inversely with the inductive reactance of the coil, since $X_L = \pi f L$ and $I = E/X_L$.

An often used type of inductive transducer is the *linear variable-differential transformer (LVDT)*. The principle of operation is illustrated in Figure 15–13B. A movable metal core is placed within an enclosure that has three windings wrapped around it. The center winding (primary) is connected to an AC source. The two outer windings have voltage induced from the primary winding. When the movable core is placed in the center of the enclosure, the voltages induced in the two

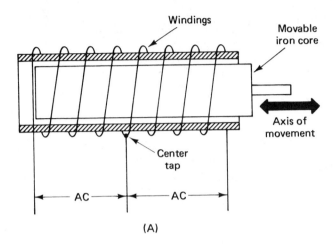

(A)

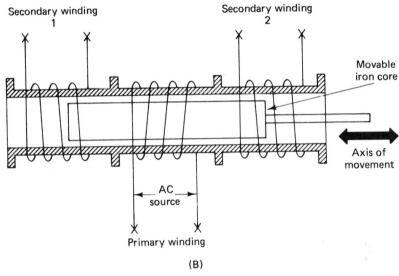

(B)

FIGURE 15–13   Inductive transducers. (A) Movable-core inductive transducer. (B) Linear variable-differential transformer.

outer windings are equal. Any movement of the core in either direction will cause one induced voltage to increase and the other induced voltage to decrease. It is possible to sense the difference in voltage induced into the two outer windings in terms of the amount of movement of the core. The variation in flux linkage due to the movement of the metal core is responsible for the change in induced voltages. Thus, a linear movement (physical quantity) can be converted to an electrical quantity by this type of inductive transducer.

PIEZOELECTRIC TRANSDUCERS

When certain crystalline materials are subjected to a mechanical stress, an electrical potential is developed across the material. Crystals such as quartz and Rochelle salt exhibit this characteristic. An application of the piezoelectric principle is the cartridge/needle assembly of a phonograph (see Figure 15–14A). The cartridge contains a crystalline material which vibrates in accordance with the variations in the grooves of a phonograph record as the needle (which is attached to the cartridge) rides through the record grooves. The crystalline material develops an electrical potential across its structure due to the mechanical vibrations. These small electrical variations are then amplified by the sound system. Thus, mechanical energy (vibrations) is converted to electric energy by piezoelectric transducers.

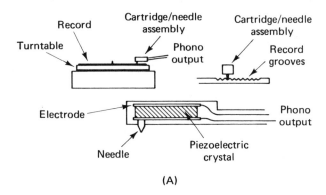

(A)

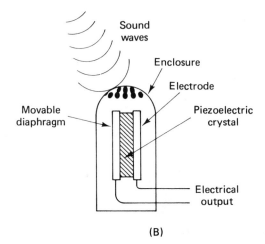

(B)

FIGURE 15–14   Piezoelectric transducers. (A) Phonograph cartridge. (B) Crystal microphone.

It is also possible to convert sound energy to electric energy with piezoelectric transducers. This is commonly done with crystal microphones (see Figure 15–14B) in which sound waves cause the vibration of a piezoelectric crystal. An electrical potential is then developed across the crystal and is amplified by the sound system.

THERMOELECTRIC TRANSDUCERS

Transducers that produce a change in electrical characteristics due to a change in temperature are referred to as *thermoelectric transducers*. Some transducers of this type, such as the *resistance thermometer*, change in resistance when the temperature changes. Others, such as the *thermocouple*, produce a voltage when heated.

The resistance thermometer is ordinarily made of a fine wire-mesh assembly of nickel, platinum, or copper wire as shown in Figure 15–15A. A wire is insulated within a thin ceramic or plastic material. If a constant DC potential is applied to the resistive element, a constant current will be indicated on a meter. However, if the resistive element is heated, its resistance will increase. The increase in resistance will cause the current reading to decrease. Due to the small-diameter wire element used, resistance thermometers respond rapidly to changes in temperature. They can also be used with bridge circuitry for making precise comparative measurements of temperature.

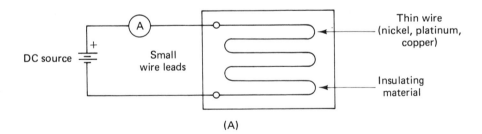

(A)

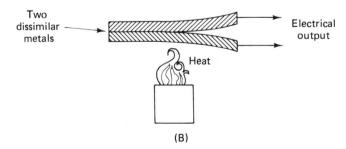

(B)

FIGURE 15–15 Thermoelectric transducers. (A) Resistance thermometer. (B) Thermocouple.

Thermovoltaic transducers are referred to as *thermocouples*. A thermocouple, illustrated in Figure 15–15B, consists of two dissimilar metals that are fused together at one end. The metals are usually combinations of iron-constantan, copper-constantan, platinum-rhodium, or other metals that exhibit the thermocouple principle. When the fused end of the two dissimilar metals is heated, a voltage will be developed at the ends that are not connected. The potential difference exists due to the different coefficients of expansion of the metals. The voltage produced by thermocouples is usually in the millivolt range. Various wire combinations are used to respond to different ranges of temperature. Thermocouples are transducers since they convert heat energy into electric energy.

OPTOELECTRIC TRANSDUCERS

An additional type of transducer that has recently received emphasis for industrial applications is the *optoelectric* or photoelectric transducer. Such transducers are light-sensitive devices that convert changes of light energy into changes of electric energy. As an example, when a photovoltaic cell is exposed to light, it will develop a voltage output. Other optoelectric transducers convert change of light into changes of resistance. Optoelectric transducers were discussed in Chapter 10.

In this chapter the characteristics of the cathode-ray tube, the thermistor, the electromagnetic and solid-state relay, and the transducer have been examined. Answer the following review questions, solve the problems, and perform the suggested laboratory activities which are designed to increase understanding of the information presented in this chapter.

REVIEW

1. What is a CRT?

2. Why is the CRT important to the area of electronics?

3. What roles do the CRT's filament and cathode play in the device's operation?

4. What is Aquadag?

5. How is the vertical and horizontal deflection of the CRT's electron beam controlled?

6. What role does phosphor play in causing the CRT's trace to be displayed?

7. What characteristics are associated with a device that exhibits a negative temperature coefficient?

8. What are the differences between a directly and indirectly heated thermistor?

9. What three characteristics are associated with the thermistor that make it useful as an electronic device?

10. What is a thermistor's resistance-temperature characteristic?

11. What is a thermistor's voltage-current characteristic?

12. What is a thermistor's current-time characteristic?

13. What is the self-heating effect associated with the operation of a thermistor?

14. What is the dissipation constant of a thermistor?

15. What is a thermistor's time constant?

16. What are two classifications of relays?

17. What is meant by the terms *make current* and *break current*?

18. How do the NO and NC contacts of an electromagnetic relay react when the relay coil picks up?

19. How do the NO and NC contacts of an electromagnetic relay react when the relay coil drops out?

20. What are some differences in the electromagnetic and solid-state relays?

21. What specifications are associated with the electromagnetic relay?

22. What specifications are associated with the solid-state relay?

23. What are some applications for relays?

24. Discuss the following types of transducers:

    a) Resistive

    b) Capacitive

    c) Inductive

    d) Piezoelectric

    e) Thermoelectric

PROBLEMS

1. The time constant of a thermistor is 0.03 s and its resistance is 5000 Ω. If it is taken from an environment of 0 °C and placed in an environment of 100 °C, what is its temperature and resistance after 1.5 s?

2. How many seconds would be required for the above thermistor to exhibit a resistance caused by 100 °C?

SUGGESTED LABORATORY ACTIVITIES

1. Measure and record the resistance of the thermistor when it is at room temperature.

2. Grasp the thermistor with your hand for a period of 3 min. Measure and record its resistance after grasping it for 3 min.

3. How does the resistance recorded in Lab 1 compare to the resistance in Lab 2?

4. How do you account for the difference?

5. Construct the circuit illustrated in Figure 15–16 and measure and record the circuit current and voltage drop across R_1 when the thermistor is at room temperature.

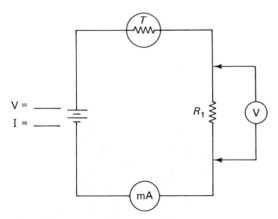

FIGURE 15–16   Thermistor circuit.

6.   Submerge the tip of the thermistor in a cup of very cold water for a period of 2 min and record the new values of I and V.

7.   How do the measurements in Lab 5 compare to the measurements in Lab 6?

8.   Construct the circuit illustrated in Figure 15–17 and measure and record the relay's pickup and dropout current.

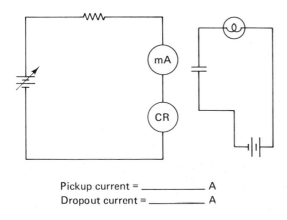

Pickup current = _____ A
Dropout current = _____ A

FIGURE 15–17   Relay circuit.

CHAPTER 16
Electronic Device Testing and Measuring Equipment

Most electronic devices may be tested very easily to determine if they are functional. Simple tests of most devices can be made with an ohmmeter. A knowledge of the operational characteristics of devices is necessary to make ohmmeter tests. In addition to ohmmeter tests, there are many types of equipment available to test electronic devices. In this chapter the testing of common electronic devices such as diodes, junction transistors, FETs, SCRs, and digital devices will be discussed. In addition several common types of measuring equipment are studied.

DIODES

Diode operation is dependent upon the voltage polarity applied to the device. Current flow occurs when it is forward biasing and very little current flows during reverse biasing. A person working with diodes must be certain that they are connected into a circuit properly.

The P material of the diode is the anode and the N material is the cathode. Forward current passes through a diode from the cathode to the anode. Diodes are available in many different package types and styles. There is often confusion about lead identification. A technician should be able to test diodes and identify leads. Volt-ohm-milliammeters (VOMs) or transistorized voltmeters (TVMs) may be used to perform this type of test. The ohmmeter section of the instrument is used to perform diode testing.

Figure 16–1 shows an ohmmeter used to test a diode. The meter should usually be placed in the $R \times 1$-kΩ range. The two meter test leads are then connected to the diode. Figure 16–1A shows an ohmmeter connected to forward bias a diode (+ to anode and − to cathode). This test should cause the meter to show a *low resistance*. Reverse biasing of a diode is shown in Figure 16–1B. An *infinite resistance* reading should be indicated by the ohmmeter during reverse biasing (+ to cathode and − to anode).

A good diode should have low resistance when forward biased and high resistance when reverse biased. A diode that has low resistance in both directions is shorted. Shorting can be caused when the maximum current or voltage ratings of the device are exceeded. *Shorted diodes* often cause damage to other circuit components.

When a diode shows infinite resistance in both directions it is open. *Open diodes* cause a circuit to be nonconductive. Open diodes do not occur very often in electronic circuits.

An ohmmeter can also be used to identify the leads of a diode. The polarity of the ohmmeter voltage source must be known. On most ohmmeters the black or common lead is negative and the red lead is positive. A diode connected to the ohmmeter will have low resistance when forward biased and infinite resistance when reverse biased. Forward biasing occurs when the polarity of the ohmmeter matches the polarity of the diode. The positive lead is then connected to the anode and the negative lead to the cathode. Lead identification is simply a matter of noting the polarity of the ohmmeter leads.

Ohmmeters with reverse polarity have a black positive lead and a red negative lead. Diode lead identification may be determined by connecting the ohmmeter to the diode in the direction of forward biasing. Forward biasing causes the ohmmeter to read a low resistance. The lead polarity of the diode is now the reverse of the ohmmeter lead polarity. The positive lead is connected to the cathode and the negative lead to the anode.

It is important that the polarity of an ohmmeter be known before it is used to identify diode leads. A meter can be checked with a DC voltmeter. Connect the voltmeter to the ohmmeter probes and observe the voltage reading. A polarity test should be made on all ranges of the ohmmeter. Some meters may have a different polarity on the higher resistance ranges.

The voltage value of an ohmmeter source is also important when testing diodes. If the supply voltage is less than about 0.2 V, it may not be large enough to forward bias a diode. This might cause a good diode to test as being open by showing an infinite resistance in both directions. It takes at least 0.3 V to cause conduction in germanium diodes and 0.7 V for silicon diodes. The ohmmeter voltage must be larger than these

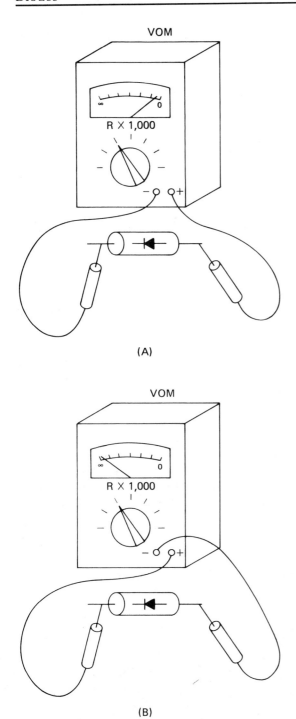

FIGURE 16–1 Diode testing. (A) Forward-bias test connection.
(B) Reverse-bias test connection.

values to test diodes. The supply voltage of ohmmeters should not be extremely large. This type of meter might damage certain diodes.

ZENER DIODES

Testing a junction diode shows that when forward biased, a current will flow through the diode. Reverse biasing will not cause current flow until a high value of reverse voltage is applied. If reverse voltage is large enough, breakdown will occur and cause a reverse current flow. Junction diodes are usually destroyed when this occurs. Zener diodes, however, are designed to operate in the reverse direction without being damaged.

Figure 16–2 shows a test circuit for a zener diode. In the forward-bias direction zener diodes behave like junction diodes. In the reverse-bias direction there is very little reverse current flow until the reverse breakdown voltage is reached, causing an increase in reverse current flow. Reverse current can pass through a zener diode without damaging it. The voltage (V_z) across the diode will remain constant. Zener diodes are available in many values of breakdown voltage and power ratings.

Zener diodes may be tested to see if they are good or bad in the same way as junction diodes. However, this test does not fully check the reverse voltage characteristic which is very important for zener diodes. The simple test circuit of Figure 16–2 may be used for an active test of zener diode characteristics. Potentiometer R_1 is adjusted to apply either forward or reverse voltage to the diode. The values of zener voltage (V_z) and zener current (I_z) may be monitored as source voltage (V_S) is varied.

When testing a zener diode, a chart similar to the one shown in Figure 16–3 should be completed. The diode should first be tested with

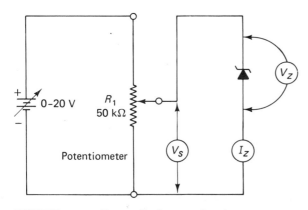

FIGURE 16–2   Zener diode test circuit.

Applied DC voltage (V_S)	V_Z	I_Z
0 V		
4 V		
8 V		
12 V		
16 V		
20 V		

(A)

Applied DC voltage (V_S)	V_Z	I_Z
0 V		
4 V		
8 V		
12 V		
16 V		
20 V		

(B)

FIGURE 16–3   Zener voltage versus zener current. (A) Forward bias. (B) Reverse bias.

an ohmmeter to see that it is not shorted or open. The test circuit should be used to adjust voltage values up to the reverse voltage rating. The diode characteristics should be similar to those shown in Figure 16–4, with the value of V_z being where conduction occurs in the reverse direction.

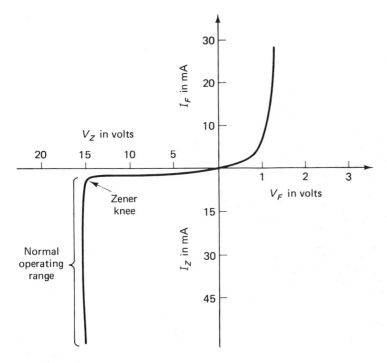

FIGURE 16–4   Zener diode characteristics.

TRANSISTORS

Transistor testing must be performed periodically when working with electronic equipment. Lead identification, gain, opens, shorts, and leakage are some of the tests commonly performed on a transistor. Testers such as the one shown in Figure 16–5 can easily be used for these tests. These instruments are not always available. An ohmmeter can be used to perform many tests.

TRANSISTOR JUNCTION TESTING

A transistor has two PN junctions which may be tested with an ohmmeter. Figure 16–6 shows the PN junctions of NPN and PNP transistors. Each junction should respond similar to a diode when it is

FIGURE 16–5   Transistor tester. (Courtesy of Sencore, Inc.)

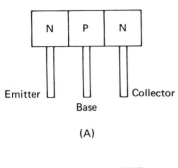

(A)

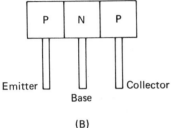

(B)

FIGURE 16–6   Transistor PN junctions. (A) NPN transistor. (B) PNP transistor.

tested. Forward biasing occurs when the polarity of the junction matches the voltage polarity of the ohmmeter. When the ohmmeter polarity is reversed, the junction will be reverse biased. Forward biasing is indicated by a low resistance reading and reverse bias by a high-resistance value. Both junctions should respond to this test in the same way if the transistor is good.

A defective PN junction will not show a resistance difference in the forward and reverse tests. Low resistance in both directions indicates that a junction is shorted. Infinite or very high resistance in both directions indicates an open condition. An open junction may be caused by a broken internal connection.

LEAKAGE TEST

A leakage test of the junctions of a transistor can also be made with the ohmmeter. Connect one ohmmeter lead to the emitter and the other to the collector. Ohmmeter lead polarity is not important for this test. This connection should cause a high resistance reading. Then reverse the two ohmmeter leads. High resistance should also occur in this direction. A good transistor will show an infinite resistance in both directions. Measurable resistance in either direction indicates *leakage*. A good transistor will have no indication of emitter-collector leakage.

LEAD IDENTIFICATION

An ohmmeter can also be used to identify the leads of a transistor. The polarity of the ohmmeter voltage source must be known in order to make this test. Most ohmmeters have a black (common) lead which is negative and a red lead which is positive. Some ohmmeters have polarities in the opposite direction. The polarity of an ohmmeter can be tested with a DC voltmeter if it is unknown.

To identify the leads, first inspect the lead location of the transistor. Assume that the center lead is the base lead. Connect the negative lead of the ohmmeter to it as shown in Figure 16–7. Then touch the positive ohmmeter lead to each of the two outside transistor leads. If a low resistance occurs for each lead, the center lead is actually the base. This test indicates that the transistor is a PNP type. If the resistance is high between the center lead and the two outside leads, reverse the meter polarity. The positive lead should be connected to the assumed base. Alternately switch the negative ohmmeter lead between the two outside transistor leads. If a low resistance reading is obtained, the transistor is an NPN type.

If the center lead does not produce low resistance in either of the two situations, it is not the base. One of the other outside leads should then be assumed as the base and the same procedure tried again. If this does not produce proper results, the other outside lead should be tried. One of the three leads will respond properly as the base if the transistor is good. If proper response cannot be obtained, the transistor is not functional. The procedure for determining emitter and collector leads is discussed in the following section.

GAIN

It is possible to test the gain and to identify the emitter and collector leads of a transistor. A 100-kΩ resistor and the ohmmeter are used for testing signal transistors. The resistor is connected to provide a base current from the ohmmeter. If a power transistor is tested, the $R \times 1$ meter range and a 1-kΩ base resistor are used.

Figure 16–8 shows how PNP and NPN transistors respond to the gain test. The internal battery of the ohmmeter is used to supply operating voltage to the transistor. For PNP transistors the positive ohmmeter lead is connected to the emitter and the negative lead to the collector. When a 100-kΩ resistor is connected between the collector and base, it should cause a base current flow. The emitter-base junction is forward biased and the collector-base is reverse biased; thus, the transistor should have low resistance. The ohmmeter should indicate a low resistance.

The gain test for an NPN transistor is also shown in Figure 16–8. For an NPN transistor the emitter is connected to the negative ohmmeter lead and the collector to the positive lead. When a 100-kΩ resistor is con-

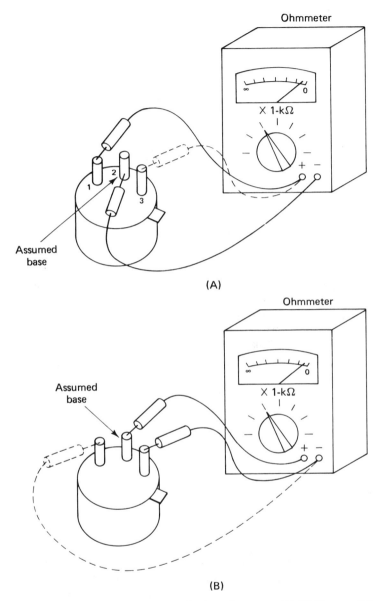

FIGURE 16–7   Transistor lead identification. (A) PNP test. (B)
NPN test.

nected between the collector and base, base current should flow. The
emitter-base junction is forward biased and the base-collector junction is
reverse biased, thus the transistor has low resistance. The meter should
now indicate low resistance.

In this circuit it was assumed that the negative ohmmeter lead is
connected to the emitter and the positive lead to the collector. If the as-

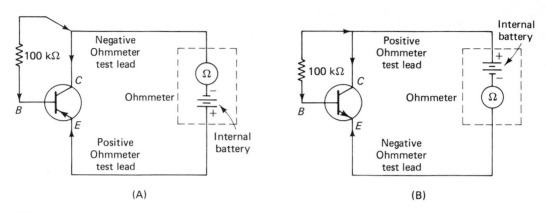

FIGURE 16–8 Transistor gain test. (A) PNP test circuit. (B) NPN test circuit.

sumed leads are correct, the emitter is forward biased and the collector is reverse biased. Connecting the base resistor to the positive lead should cause a base current flow. The ohmmeter should then indicate a low resistance. The emitter-base junction is forward biased and the base-collector junction is reverse biased. If no base current flows, the assumed emitter-collector leads should be reversed. Reverse the ohmmeter leads and repeat the procedure. If the transistor is good and the assumed leads are correct, the ohmmeter will indicate a low resistance. If not, the transistor has low gain.

JFETs

JFET testing with an ohmmeter is easy to accomplish. The JFET is a single bar of silicon with one PN junction. To identify the leads, select any two of the three leads and connect the ohmmeter to these leads. If the resistance is 100–1000 Ω, the two leads are the source and drain. Reverse the ohmmeter leads and connect them to the same leads. If the two leads are actually the source and drain, the *same* resistance will be indicated. Thus, the remaining lead is the gate. If the two leads show a different resistance in each direction, one is the gate. Repeat the procedure until the JFET shows the same resistance in each direction between the two leads. These are the source and drain leads of the JFET.

The gate of a JFET is low-resistant in one direction and high-resistant in the reverse direction. If the JFET shows low resistance when the positive probe is connected to the gate and negative to the source or drain, the JFET is an N channel type. A P channel device shows low resistance when the gate is negative and the source or drain is positive.

The resistance ratio in the forward and reverse direction should be at least 1 : 1K for a good JFET. Figure 16–9 shows some typical resistance values for JFETs.

A test of JFET amplification is accomplished by connecting an ohmmeter between the source and drain. The polarity of the meter is not important for this test. When the gate of the JFET is touched, a decrease in resistance should occur. Even more pronounced resistance change should take place when a 200-kΩ resistor is placed momentarily between the gate and either the source or drain. If very little or no resistance change occurs, the JFET is not functional.

D-MOSFETs

D-MOSFETs may be tested using methods similar to JFET testing. The gate of a D-MOSFET is insulated from the channel. A D-MOSFET should have infinite resistance between each lead regardless of ohmmeter polarity. Some typical resistance values of D-MOSFETs are shown in Figure 16–10. Three-lead D-MOSFETs have the substrate and source internally connected, while four-lead D-MOSFETs have a separate substrate lead.

A four-lead D-MOSFET is either an N channel or a P channel device. A P channel device has low resistance when the positive lead is connected to the substrate and negative lead to either the source or drain. An N channel device has low resistance when the negative lead is connected to the substrate and the positive lead to either the source or drain.

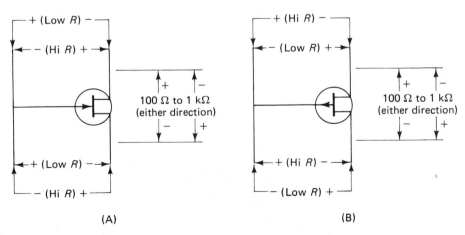

(A)   (B)

FIGURE 16–9   Typical JFET resistances. (A) N channel. (B) P channel.

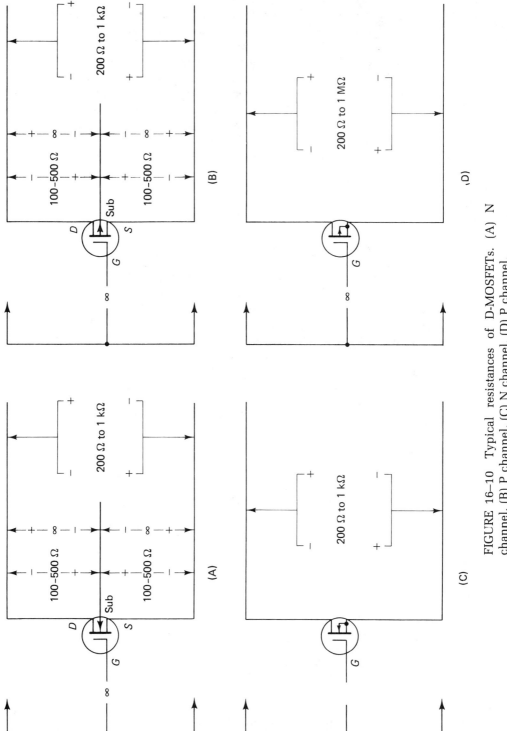

FIGURE 16–10 Typical resistances of D-MOSFETs. (A) N channel. (B) P channel. (C) N channel. (D) P channel.

E-MOSFETs

E-MOSFET testing with an ohmmeter is not an accurate test. The structure of this device does not allow an operational check of the device. The gate, source, and drain of a good device should have infinite resistance between each of the leads. A four-lead device should respond in the same way as a diode between the substrate and the source or drain. Ohmmeter polarity can be used to identify the substrate and the channel. A three-lead device should respond as a diode between the source and drain. Figure 16–11 shows some typical ohmmeter resistances of E-MOSFETs.

UJTs

UJT testing with an ohmmeter is identical to testing JFETs. However, the resistance between B_1 and B_2 should be in the range of 1–10 kΩ. For most UJTs the resistance from B_2 to the emitter is somewhat less than from B_1 to emitter. UJTs should not change resistance from B_1 to B_2 when the emitter is touched.

SCRs

A VOM may be used to test SCRs. The $R \times 1$ resistance range is ordinarily used. Digital meters (DVMs) do not normally perform this test well because they have low current capabilities in the ohmmeter ranges. Any two SCR leads may be selected to test the forward and reverse resistance. If there is a resistance in both directions, the two leads are the cathode and gate and the third lead is the anode. If the resistance is infinite in both directions, change one of the two leads. Test again for a resistance indication in one direction. The position of the anode lead is important. SCRs should have the lowest resistance reading when the gate is positive and the cathode is negative.

To complete the SCR test, connect the negative ohmmeter lead to the cathode and the positive lead to the anode. Momentarily touch the gate to the positive anode lead. This should cause a low resistance reading on the ohmmeter. The gate should then be disconnected from the anode lead. The ohmmeter should continue to indicate low resistance, showing the switching function of the SCR. If the SCR responds properly, the SCR is functional.

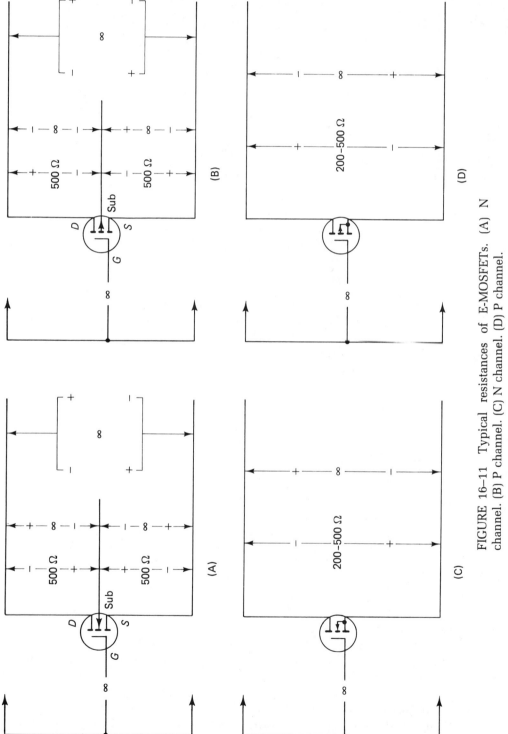

FIGURE 16–11 Typical resistances of E-MOSFETs. (A) N channel. (B) P channel. (C) N channel. (D) P channel.

TRIACs

Triacs may also be tested with a VOM, using the $R \times 1$ range. DVMs and TVMs usually do not work well for this test since their current capability is lower than VOMs. Any two leads are selected at random and the forward and reverse resistance is measured. T_1 and T_2 should have infinite or very high resistance in either direction. If this does not occur, select another lead and check the resistance ratio until T_1 and T_2 are located. The third lead is the gate terminal. Use either the T_1 or T_2 lead and the gate and measure the forward and reverse resistance. T_1 to gate should have low resistance in both the forward and reverse direction. T_2 to gate should have an infinite resistance in both directions.

To test the *latching* condition, connect the positive ohmmeter lead to T_2 and the negative lead to T_1. Momentarily touching the gate to T_2 should cause a low resistance reading. Disconnecting the gate should cause resistance between T_1 and T_2 to remain the same. Connecting the positive ohmmeter lead to T_1 and the negative lead to T_2 should cause a similar response. Momentarily touching the gate to T_2 should cause a low resistance reading. Disconnecting the gate should cause a low resistance reading to remain between T_1 and T_2. This shows that the triac responds as an AC switch and will latch in either direction.

DIGITAL CIRCUITS

The use of digital circuits for computer applications has added a new dimension to electronic testing. Many of the measurements associated with digital circuitry are tests for 1 and 0 logic levels. These logic levels are indicated by either the presence or absence of voltage. Such tests may be made by using electronic meters or by *logic probes* such as the one shown in Figure 16–12.

Another digital circuit testing device is the *logic clip* shown in Figure 16–13. A logic clip may be used to very rapidly check input and output conditions of integrated circuit chips. The indicating lamps at the top of the logic clip show either a voltage or no voltage. This gives a quick indication of the 1 and 0 states of digital circuits.

MEASURING EQUIPMENT

There are many types of measuring equipment available today for testing electronic devices and testing circuits. Several types of electronic measuring equipment are discussed in the following sections. The most basic

FIGURE 16–12 Digital logic probe. (Courtesy of Hewlett-Packard Corp.)

FIGURE 16–13 Digital logic clip. (Courtesy of Hewlett-Packard Corp.)

types are mentioned first, with some more sophisticated equipment following.

HAND-DEFLECTION METERS

Meters that rely on the motion of a hand or pointer are called *hand-deflection meters*. The volt-ohmmeter (VOM) is a common type of hand-deflection meter. Figure 16–14 shows some hand-deflection VOMs. VOMs are used to measure several different electrical quantities. *Single-function meters* may also be used to measure electrical quantities. They measure only one quantity, while *multifunction meters*, like VOMs, may be used to measure several electrical quantities.

METER MOVEMENTS

The basic part of a hand-deflection meter is called the *meter movement*. The movement of the hand or pointer over a calibrated scale indicates the quantity being measured.

Many meters use the *D'Arsonval* or *moving-coil* type of meter movement. The construction details of this meter movement are shown in Figure 16–15. The hand or pointer of the movement is positioned on the left side of the calibrated scale. A moving coil is located inside a horseshoe magnet. Current flows through the coil from the circuit being tested. A reaction occurs between the electromagnetic field of the coil

FIGURE 16–14   Electric meters—hand-deflection type. (Courtesy of Triplett Corp.)

FIGURE 16–14 *(cont.)*

and the permanent magnetic field of the horseshoe magnet. This reaction causes the hand to move toward the right side of the scale. This movement of the coil is proportional to the current flow. This moving-coil meter movement operates on the same principle as an electric motor by producing a torque due to the interaction of two magnetic fields. Moving coils can be used for single-function meters or for multifunction meters.

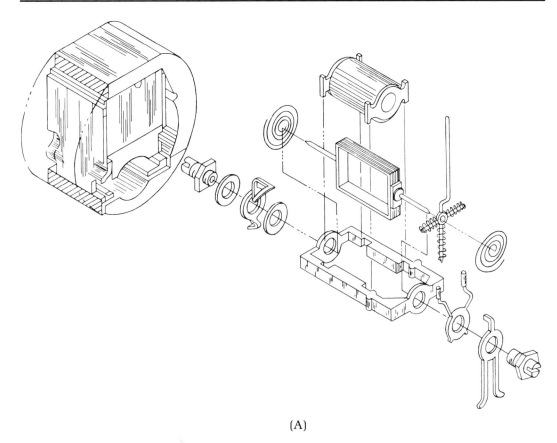

(A)

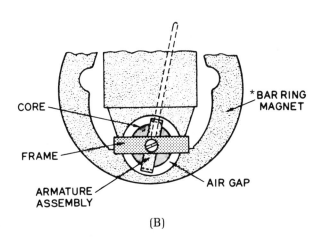

(B)

FIGURE 16–15 Construction details of moving-coil meter movement. Courtesy of Triplett Corp. (A) Exploded view of DC pivot- and jewel-type meter. (B) Partial view showing air gap.

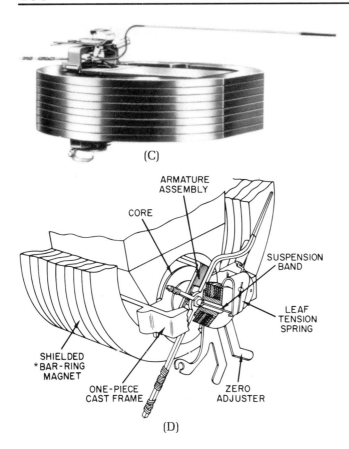

(C)

ARMATURE
ASSEMBLY

CORE

SUSPENSION
BAND

LEAF
TENSION
SPRING

SHIELDED
*BAR-RING
MAGNET

ONE-PIECE
CAST FRAME

ZERO
ADJUSTER

(D)

FIGURE 16–15 *(cont.)*   (C) Suspension movement. (D) Suspension drawing.

A meter movement may be used to measure voltage, current, resistance, or other quantities. It is necessary to *calibrate* the meter scale to accurately measure the desired quantity.

METER SENSITIVITY

An important characteristic of meters is *sensitivity*. Sensitivity increases as the current rating (I_m) decreases. Voltmeters are connected in parallel with a circuit to measure voltage. Part of the current flows through the meter to make the needle deflect. The current that flows through the meter *should* be very small. Meters which draw excessive current from the circuit are said to "load" the circuit. Meter loading of circuits being tested must be reduced to a minimum. More sensitive meters draw less current from the circuit being tested. Sensitivity is measured in ohms per volt and is equal to $1/I_m$. I_m is the current required for full-scale deflection of the meter. A 100-μA movement has an *ohm-per-volt rating* of 1/0.0001 or 10,000 Ω/V. This means that the meter has 10,000 Ω of resistance for its 1-V range or 100 kΩ for the 10-V range. Sensitivities of meter movements range from as low as 100 Ω/V to as high as 200,000 Ω/V. Electronic meters such as digital voltmeters have sensitivities in excess of 1 MΩ/V. Low-sensitivity meters should not be used for making precise measurements. An example of the importance of meter sensitivity is shown in Figure 16–16.

MEASURING AC VOLTAGE AND CURRENT

The moving-coil meter movement may be modified to indicate values of AC voltage or current. Either half-wave or full-wave rectifier circuits are used to convert alternating current from an external circuit to

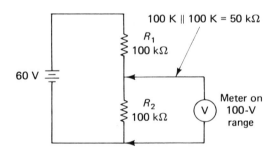

(A)

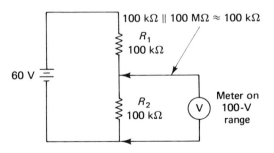

(B)

FIGURE 16–16 Examples of meter sensitivity. (A) Using a 1000-Ω/V sensitivity value. (B) Using a 1-MΩ/V sensitivity value.

direct current that will produce an electromagnetic field in the coil of the meter movement. Half-wave and full-wave rectifier circuits for meters are shown in Figure 16–17. These types of rectifiers are used to convert alternating current to direct current to cause the pointer of the movement to deflect.

MEASURING ELECTRIC POWER

Electric power is measured with a *wattmeter*, which uses a meter movement called a *dynamometer* (see Figure 16–18). The dynamometer movement has two electromagnetic coils. One coil, called the *current coil*, is connected in series with the circuit to be measured. The other coil, called the *potential coil*, is connected in parallel with the circuit. The strength of both electromagnetic fields affects the amount of movement of the meter pointer. The operational principle of this movement is similar to the moving-coil type, except it uses an electromagnetic field rather than a permanent magnetic field. The interaction of the two electromagnetic fields produces a torque which moves the meter pointer.

When measuring DC power, total power is the product of voltage and current $(P = E \times I)$. However, when measuring AC power the *power factor* of the load must be considered. The *true power* of an AC circuit is measured with a wattmeter.

COMPARATIVE METERS

Comparative meters are designed to compare a *known value* to *some unknown value*. The accuracy of comparative meters is much greater than that of hand-deflection meters. They may be used to make very accurate measurements. The *Wheatstone bridge*, shown in Figure 16–19, is a comparative meter. A voltage source is used along with a sensitive zero-centered meter movement and a resistive bridge circuit. The bridge circuit is completed by adding an unknown external resistance (R_x) as the resistance to be measured. The known resistance (R_s) is adjusted so that the resistive path formed by R_x and R_s is equal to the path formed by R_A and R_B. Thus, no current flows through the meter. In this condition, the meter indicates zero. This is referred to as a *null* and the bridge is said to be *balanced*. The value of R_s is marked on the meter so that the value of R_x can be determined by the adjustment. Resistors R_1 and R_2 form the *ratio arm* of the circuit. The value of the unknown resistance (R_x) is found by the equation

$$R_x = \frac{R_A}{R_B} \times R_s$$

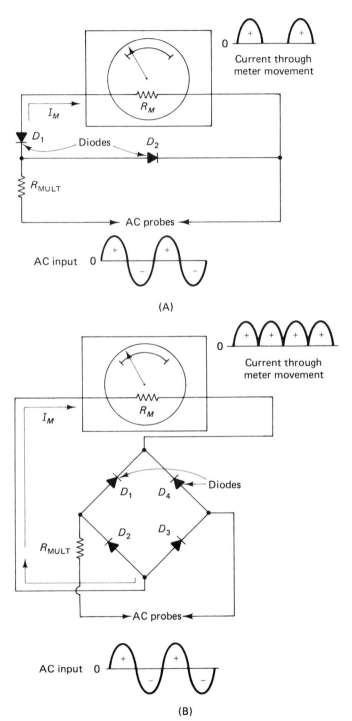

FIGURE 16–17 Measuring alternating current. (A) Half-wave
rectifier circuit. (B) Full-wave rectifier circuit.

The Moving Iron-Vane Movements are used to measure RMS AC voltages and currents.

A. Moving Iron-Vane Movements are fluid-damped to control the meter dynamic characteristics.

B. The mounting plate is one molded piece which reduces the number of parts.

C. Field coils are molded for greater strength and dependability.

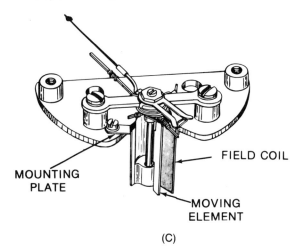

(C)

FIGURE 16–17 (cont.) (C) AC moving iron-vane type meter movement (Courtesy of Triplett Corp.).

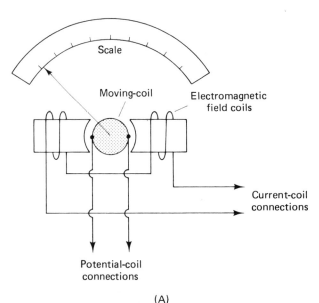

(A)

FIGURE 16–18 Dynamometer meter movement to measure power. (A) Dynamometer movement diagram. (B) Dynamometer wattmeter circuit.

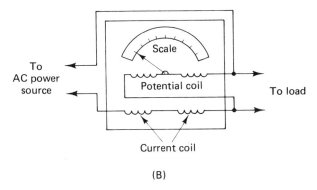

(B)

Dynamometers are used to measure AC and DC power.

A. Dynamometer movement consists of two coils. One coil is stationary while the other coil is movable and is affixed to the moving element (pointer, etc.).

B. The moving element is fluid-damped to control the dynamic characteristics.

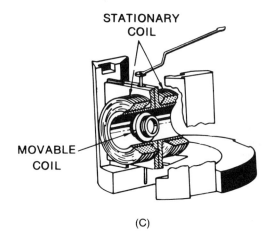

STATIONARY COIL

MOVABLE COIL

(C)

FIGURE 16–18 (cont.) (B) Dynamometer wattmeter circuit. (C) Dynamometer movement (Courtesy of Triplett Corp.).

A Wheatstone bridge is capable of measuring values of resistance from 0.001 Ω to 10 MΩ with considerable accuracy. Many other types of comparative meters use the Wheatstone bridge principle. Comparative meters compare an unknown quantity with a known quantity contained within the indicator.

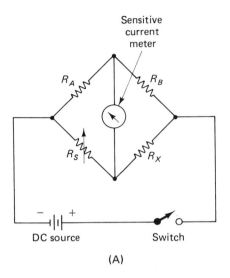

(A)

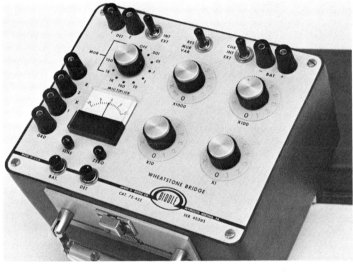

(B)

FIGURE 16–19   Wheatstone bridge—type of comparative meter. (A) Circuit diagram. (B) Courtesy of Biddle Instruments.

Another technique used with comparative meters is the *potentiometric* method. Bridges are used to measure impedance while the potentiometric method is used to measure voltages accurately. The illustration
of Figure 16–20 shows the simplified potentiometric method. The voltage to be measured is applied to the input terminals of the circuit. The
potentiometer, which is connected across the reference supply voltage,
forms a voltage divider circuit. When the movable arm of the potentiometer is adjusted to a voltage (V_{ref}) equal to the voltage being measured,

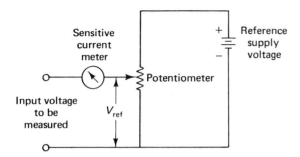

FIGURE 16–20   Potentiometric meter.

the current through the meter is zero. When the zero (or null) reading on the zero-centered meter is obtained, the value of the unknown voltage can be read from a calibrated scale on the potentiometer.

Since no current is drawn from the voltage source being measured during the null condition, the meter can be considered an infinite impedance. Thus, the potentiometric technique can be used for precise calibration of other meters. Potentiometric meters are also used as laboratory voltage standards.

OSCILLOSCOPES

The cathode-ray tube (CRT) *oscilloscope* is an important type of measuring instrument (see Figure 16–21). It is possible to monitor the voltage of a circuit visually by using an oscilloscope. The basic operational part of an oscilloscope is the CRT. Figure 16–22 illustrates the electron gun assembly in the neck of the CRT. A beam of electrons is produced by the cathode of the tube when it is heated by a filament voltage. Electrons are attracted to the positive potential of anode 1 and move in that direction. The quantity of electrons passing on to anode 1 is determined by the amount of negative bias voltage applied to the control grid. Anode 2 is operated at a higher positive potential than anode 1 to further accelerate the electron beam toward the screen of the CRT. The difference in potential between anode 1 and anode 2 determines the point of convergence of the beam on the CRT screen. When the electron beam strikes the phosphorescent screen of the CRT, light is emitted. Thus, a visual image or trace is produced on the screen.

To control the horizontal and vertical movement of the electron beam, deflection plates are used. If no potential is applied to either plate, the electron beam would merely appear as a dot in the center of the CRT screen. Electrostatic deflection occurs when a potential is

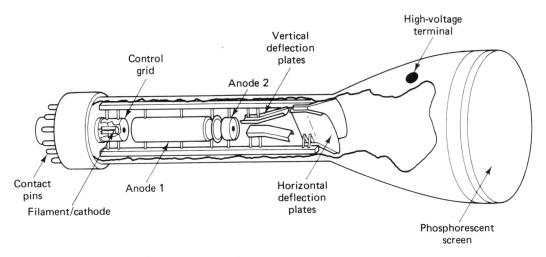

FIGURE 16–21   Internal construction of CRT.

placed on the horizontal and vertical deflection plates. Horizontal deflection usually results from a sawtooth waveform produced by a sweep oscillator circuit (see Figure 16–23), which is part of the oscilloscope internal circuitry. This voltage "sweeps" the electron beam back and forth across the CRT screen. The sweep setting of a CRT is adjusted to coincide with the frequency range of the voltage being measured. Vertical deflection takes place in accordance with the magnitude of the applied

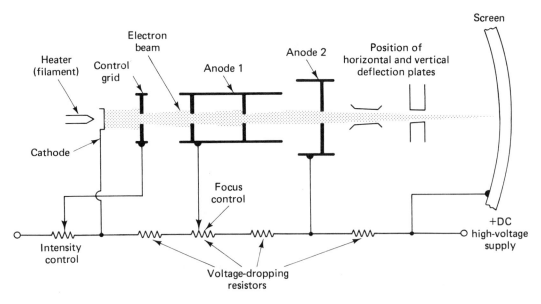

FIGURE 16–22   Electron gun assembly of an oscilloscope CRT.

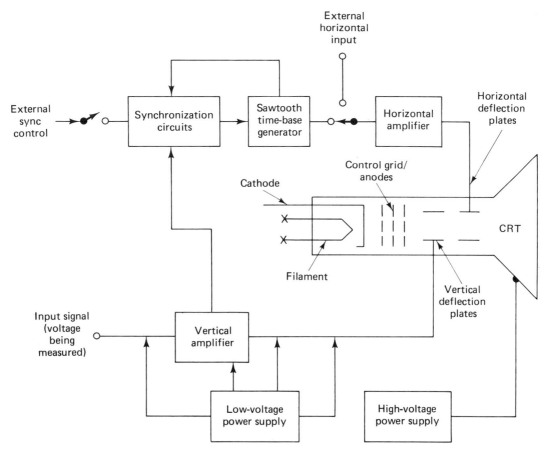

FIGURE 16–23 Block diagram of internal circuitry of oscilloscope.

voltage. The voltage to be measured is applied to the vertical deflection circuits of the oscilloscope.

Oscilloscopes are available in several different types. General-purpose oscilloscopes are used to display simple types of waveforms and for general electronic servicing. Triggered-sweep oscilloscopes are used where it is desired to apply external voltage to the horizontal circuits to produce horizontal sweep for more accurate measurements. Other oscilloscopes, classified as laboratory types, have very high sensitivity and frequency response over a wide range. Oscilloscopes are used to measure AC and DC voltages, frequency and phase relationships, distortion in amplifiers, and various timing and numerical-control applications. Memory and storage-type oscilloscopes are also available for more sophisticated measurement applications.

Oscilloscopes provide a visual display of waveforms on their screen and may be used to measure a wide range of AC frequencies. Oscilloscopes, or "scopes," are used to examine waveshapes, particularly for electronic servicing and troubleshooting.

Oscilloscopes permit voltage waveforms to be visually analyzed by producing an image on a screen. The image, called a *trace*, is usually a line on the CRT screen. A stream of electrons strikes the phosphorescent coating on the inside of the screen, causing the screen to emit light.

Oscilloscopes display voltage waveforms on two axes, similar to a graph. The horizontal axis on the screen is the *time* axis and the vertical axis is the *voltage* axis. For the CRT to display a trace properly, the internal circuits of the scope must be properly adjusted by controls on the front of the oscilloscope. Many oscilloscopes have the following controls:

1.   *Intensity*. Controls the brightness of the trace; sometimes the on-off control also.

2.   *Focus*. Adjusts the thickness of the trace.

3.   *Vertical position*. Adjusts the entire trace vertically.

4.   *Horizontal position*. Adjusts the entire trace horizontally.

5.   *Vertical gain*. Controls the vertical height of the trace.

6.   *Horizontal gain*. Controls the horizontal size of the trace.

7.   *Vertical attenuation*. Acts as a "coarse" adjustment to reduce the trace vertically.

8.   *Horizontal sweep*. Controls the speed at which the trace moves across (sweeps) the CRT horizontally and determines the number of waveforms displayed on the screen.

9.   *Synchronization select*. Controls how the input to the scope is "locked in" with the circuitry of the scope.

10.   *Vertical input*. External connections used to apply an input to the vertical circuits of the scope.

11.   *Horizontal input*. External connections used to apply an input to the horizontal circuits of the scope.

FREQUENCY MEASUREMENT

Another type of measurement is frequency. Frequency refers to the number of cycles of voltage or current that occurs in a given period of time. The international unit of measurement for frequency is the hertz (Hz), which is defined as one cycle per second. A table of frequency bands is shown in Figure 16–24. There are many frequency ranges, as shown in the figure.

Frequency can be measured with several different types of instruments. An electronic counter is one type of frequency meter. Vibrating-reed frequency meters are also common for measuring power frequencies. An oscilloscope can also be used to measure frequency.

A common method of frequency measurement for sine-wave voltages is a comparative method which uses the oscilloscope. This technique relies upon *"Lissajous" patterns* on the oscilloscope screen. An unknown frequency may be applied to the vertical input of the oscilloscope while a known value of frequency is applied to the horizontal input. The shape of the pattern which appears on the screen of the oscilloscope is used to determine the ratio of the vertical frequency to the horizontal frequency. When this ratio has been determined, the un-

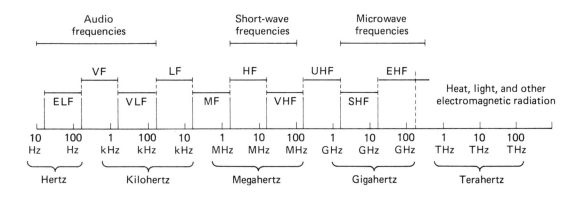

Band		Frequency range		
Extremely low frequency	(ELF)	30 Hz	TO	300 Hz
Voice frequency	(VF)	300 Hz	TO	3 kHz
Very low frequency	(VLF)	3 kHz	TO	30 kHz
Low frequency	(LF)	30 kHz	TO	300 kHz
Medium frequency	(MF)	300 kHz	TO	3 MHz
High frequency	(HF)	3 MHz	TO	30 MHz
Very high frequency	(VHF)	30 MHz	TO	300 MHz
Ultra high frequency	(UHF)	300 MHz	TO	3 GHz
Super high frequency	(SHF)	3 GHz	TO	30 GHz
Extremely high frequency	(EHF)	30 GHz	TO	300 GHz

FIGURE 16–24 Table of frequency bands.

known frequency can be easily calculated. The formula used to find the unknown frequency is

$$F_x = F \times \frac{T_H}{T_V}$$

where:

      F_x is the unknown frequency,

      F is the known frequency,

      T_H is the number of times the Lissajous pattern touches a horizontal line on the oscilloscope screen, and

      T_V is the number of times the pattern touches a vertical line on the screen.

      Some examples are shown in Figure 16–25. Note that one horizontal line and one vertical line are drawn tangent to the pattern.

DIGITAL METERS

      Many digital meters are now in use. They employ numerical readouts to simplify the measurement process and to make more accurate measurements. Instruments such as digital counters, digital multimeters, and

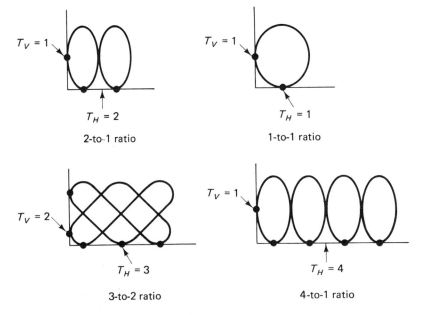

FIGURE 16–25   Lissajous oscilloscope patterns.

digital voltmeters are commonly used. Digital meters such as the one shown in Figure 16–26 rely on the operation of digital circuitry to produce a numerical readout of the measured quantity.

The readout of a digital meter is designed to transform electric signals into numerical data. Both letter and number readouts are available, as are seven-segment, discrete-number, and bar matrix displays. Each

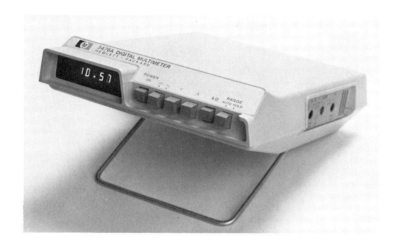

(A)

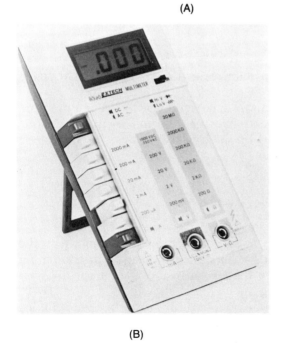

(B)

FIGURE 16–26   Digital multimeters. (A) Courtesy of Hewlett-Packard Corp. (B) Courtesy of Extech International Corp.

(C)

FIGURE 16–26 (cont.)   (C) Courtesy of Triplett Corp.

method has a device designed to change electric energy into light energy on the display.

The display of a digital meter is achieved by several electronic processes. These include the ionization of gas, heated incandescent elements, light-emitting diodes, and liquid crystal displays.

CHART-RECORDING METERS

The meters discussed previously are used where no permanent record of the measured quantity is needed. Meters are also used to provide a permanent record of some quantity that is measured over a specific time period. Typical chart-recording meters are shown in Figure 16–27. Types of chart recorders include pen-and-ink recorders and inkless recorders.

Pen-and-ink recorders use a pen attached to the meter, which touches a paper chart and leaves a permanent record of the measured

(A)

(B)

FIGURE 16–27   Chart recorders. (A) Circular chart with one recording pen—temperature recorder (Courtesy of Dickson Co.); (B) Circular chart with two recording pens—temperature and humidity recorder (Courtesy of Dickson Co.). (C) Roll chart recorder with three recording pens (Courtesy of Chessell Corp.).

(C)

FIGURE 16–27 (cont.) (C) Roll chart recorder with three recording pens (Courtesy of Chessell Corp.).

quantity on the chart. The charts utilized can either be roll charts that
revolve on rollers under the pen mechanism or circular charts that revolve on an axis under the pen (see Figure 16–27). Chart recorders may
use more than one pen to record several different quantities simultaneously (see Figure 16–27B).

The pen of a chart recorder is a capillary tube device that is actually an extension of the basic meter movement. The pen must be connected to a constant source of ink. The pen is moved by the torque ex-
erted by the meter movement just as the pointer of a hand-deflection
meter is moved. The chart used for recording the measured quantity
usually contains lines that correspond to the radius of the pen movement. Increments on the chart are marked according to time intervals.
The chart must be moved under the pen at a constant speed. Either a
spring drive mechanism, a synchronous AC motor, or a DC servomotor
can be used to record the quantities.

Inkless recorders may use a voltage applied to the pen point to produce an impression on a sensitive paper chart. In another process the
pen is heated to cause a trace to be melted along the chart paper. The
obvious advantage of inkless recorders is that ink is not required.

In this chapter electronic device testing and measuring equipment were discussed. Answer the following review questions to increase understanding of the information presented in this chapter.

REVIEW

1. Explain the procedures for testing the following devices with an ohmmeter:

    a) diode

    b) transistor

    c) zener diode

    d) JFET

    e) D-MOSFET

    f) E-MOSFET

    g) SCR

    h) triac

2. How may the leads of a diode be identified?

3. How may the leads of a transistor be identified?

4. How may the leads of an SCR be identified?

5. How is a digital logic probe used?

6. List some types of meters that are used and the quantities they measure.

7. What is a hand-deflection meter?

8. Describe the construction of a D'Arsonval meter movement.

9. What is meant by calibrating a meter scale?

10. How is meter sensitivity determined?

11. What is meant by meter loading effect?

12. How are AC voltages and currents measured with a meter movement?

13. How is electric power measured?

14. Describe the construction of a dynamometer movement.

15. What is a comparative meter?

16. What is meant by potentiometric comparison?

17. What is the major advantage of using an oscilloscope?

18. Describe the construction of a cathode-ray tube (CRT).

19. How may frequency be measured?

20. What is the major advantage of using digital meters?

21. Why are chart recorders used for measurement?

22. What are some types of chart recorders?

APPENDIX 1
Table of Common Logarithms

N	0	1	2	3	4	5	6	7	8	9
10	0000	0043	0086	0128	0170	0212	0253	0294	0334	0374
11	0414	0453	0492	0531	0569	0607	0645	0682	0719	0755
12	0792	0828	0864	0899	0934	0969	1004	1038	1072	1106
13	1139	1173	1206	1239	1271	1303	1335	1367	1399	1430
14	1461	1492	1523	1553	1584	1614	1644	1673	1703	1732
15	1761	1790	1818	1847	1875	1903	1931	1959	1987	2014
16	2041	2068	2095	2122	2148	2175	2201	2227	2253	2279
17	2304	2330	2355	2380	2405	2430	2455	2480	2504	2529
18	2553	2577	2601	2625	2648	2672	2695	2718	2742	2765
19	2788	2810	2833	2856	2878	2900	2923	2945	2967	2989
20	3010	3032	3054	3075	3096	3118	3139	3160	3181	3201
21	3222	3243	3263	3284	3304	3324	3345	3365	3385	3404
22	3424	3444	3464	3483	3502	3522	3541	3560	3579	3598
23	3617	3636	3655	3674	3692	3711	3729	3747	3766	3784
24	3802	3820	3838	3856	3874	3892	3909	3927	3945	3962
25	3979	3997	4014	4031	4048	4065	4082	4099	4116	4133
26	4150	4166	4383	4200	4216	4232	4249	4265	4281	4298
27	4314	4330	4346	4362	4378	4393	4409	4425	4440	4456
28	4472	4487	4502	4518	4533	4548	4564	4579	4594	4609
29	4624	4639	4654	4669	4683	4698	4713	4728	4742	4757
30	4771	4786	4800	4814	4829	4843	4857	4871	4866	4900
31	4914	4928	4942	4955	4969	4983	4997	5011	5024	5038
32	5051	5065	5079	5092	5105	5119	5132	5145	5159	5172
33	5185	5198	5211	5224	5237	5250	5263	5276	5289	5302
34	5315	5328	5340	5353	5366	5378	5391	5403	5416	5428

Table of Common Logarithms (Cont.)

N	0	1	2	3	4	5	6	7	8	9
35	5441	5453	5465	5478	5490	5502	5514	5527	5539	5551
36	5563	5575	5587	5599	5611	5623	5635	5647	5658	5670
37	5682	5694	5705	5717	5729	5740	5752	5763	5775	5786
38	5798	5809	5821	5832	5843	5855	5866	5877	5888	5899
39	5911	5922	5933	5944	5955	5966	5977	5988	5999	6010
40	6021	6031	6042	6053	6064	6075	6085	6096	6107	6117
41	6128	6138	6149	6160	6170	6180	6191	6201	6212	6222
42	6232	6243	6253	6263	6274	6284	6294	6304	6314	6325
43	6335	6345	6355	6365	6375	6385	6395	6405	6415	6425
44	6435	6444	6454	6464	6474	6484	6493	6503	6513	6522
45	6532	6542	6551	6561	6571	6580	6590	6599	6609	6618
46	6628	6637	6646	6656	6665	6675	6684	6693	6702	6712
47	6721	6730	6739	6749	6758	6767	6776	6785	6794	6803
48	6812	6821	6830	6839	6848	6857	6886	6875	6884	6893
49	6902	6911	6920	6928	6937	6946	6955	6964	6972	6981
50	6990	6998	7007	7016	7024	7033	7042	7050	7059	7067
51	7076	7034	7093	7101	7110	7118	7126	7135	7143	7152
52	7160	7168	7177	7185	7193	7202	7210	7218	7226	7235
53	7243	7251	7259	7267	7275	7284	7292	7300	7308	7316
54	7324	7332	7340	7348	7356	7364	7372	7380	7388	7396
55	7404	7412	7419	7427	7435	7443	7451	7459	7466	7474
56	7482	7490	7497	7505	7513	7520	7528	7536	7543	7551
57	7559	7566	7574	7582	7589	7597	7604	7612	7619	7627
58	7634	7642	7649	7657	7664	7672	7679	7686	7694	7701
59	7709	7716	7723	7731	7738	7745	7752	7760	7767	7774
60	7782	7789	7796	7803	7810	7818	7825	7832	7839	7846
61	7853	7860	7868	7875	7882	7889	7896	7903	7910	7917
62	7924	7931	7938	7945	7952	7959	7966	7973	7980	7987
63	7993	8000	8007	8014	8021	8028	8035	8041	8048	8055
64	8062	8069	8075	8082	8089	8096	8102	8109	8116	8122
65	8129	8136	8142	8149	8156	8162	8169	8176	8182	8189
66	8195	8202	8209	8215	8222	8228	8235	8241	8248	8254
67	8261	8267	8274	8280	8287	8293	8299	8306	8312	8319
68	8325	8331	8338	8344	8351	8357	8363	8370	8376	8382
69	8388	8395	8401	8407	8414	8420	8426	8432	8439	8445
70	8451	8457	8463	8470	8476	8482	8488	8494	8500	8506
71	8513	8519	8525	8531	8537	8543	8549	8555	8561	8567
72	8573	8579	8585	8591	8597	8603	8609	8615	8621	8627
73	8633	8639	8645	8651	8657	8663	8669	8675	8681	8686
74	8692	8698	8704	8710	8716	8722	8727	8733	8739	8745

Table of Common Logarithms (Cont.)

N	0	1	2	3	4	5	6	7	8	9
75	8751	8756	8762	8768	8774	8779	8785	8791	8797	8802
76	8808	8814	8820	8825	8831	8837	8842	8848	8854	8859
77	8865	8871	8876	8882	8887	8893	8899	8904	8910	8915
78	8921	8927	8932	8938	8943	8949	8954	8960	8965	8971
79	8976	8982	8987	8993	8998	9004	9009	9015	9020	9025
80	9031	9036	9042	9047	9053	9058	9063	9069	9074	9079
81	9085	9090	9096	9101	9106	9112	9117	9122	9128	9133
82	9138	9143	9149	9154	9159	9165	9170	9175	9180	9186
83	9191	9196	9201	9206	9212	9217	9222	9227	9232	9238
84	9243	9248	9253	9258	9263	9269	9274	9279	9284	9289
85	9294	9299	9304	9309	9315	9320	9325	9330	9335	9340
86	9345	9350	9355	9360	9365	9370	9375	9380	9385	9390
87	9395	9400	9405	9410	9415	9420	9425	9430	9435	9440
88	9445	9450	9455	9460	9465	9469	9474	9479	9484	9489
89	9494	9499	9504	9509	9513	9518	9523	9528	9533	9538
90	9542	9547	9552	9557	9562	9566	9571	9576	9581	9586
91	9590	9595	9600	9605	9609	9614	9619	9624	9628	9633
92	9638	9643	9647	9652	9657	9661	9666	9671	9675	9680
93	9685	9689	9694	9699	9703	9708	9713	9717	9722	9727
94	9731	9736	9741	9745	9750	9754	9759	9763	9768	9773
95	9777	9782	9786	9791	9795	9800	9805	9809	9814	9818
96	9823	9827	9832	9836	9841	9845	9850	9854	9859	9863
97	9868	9872	9877	9881	9886	9890	9894	9899	9903	9908
98	9912	9917	9921	9926	9930	9934	9939	9943	9948	9952
99	9956	9961	9965	9969	9974	9978	9983	9987	9991	9996

APPENDIX 2
Decibel Table

dB	Current or voltage ratio		Power ratio		dB	Current or voltage ratio		Power ratio	
	Gain	Loss	Gain	Loss		Gain	Loss	Gain	Loss
0	1.000	1.0000	1.000	1.0000	3.0	1.413	.7079	1.995	.5012
0.1	1.012	.9886	1.023	.9772	3.1	1.429	.6998	2.042	.4898
0.2	1.023	.9772	1.047	.9550	3.2	1.445	.6918	2.089	.4786
0.3	1.035	.9661	1.072	.9333	3.3	1.462	.6839	2.138	.4677
0.4	1.047	.9550	1.096	.9120	3.4	1.479	.6761	2.188	.4571
0.5	1.059	.9441	1.122	.8913	3.5	1.496	.6683	2.239	.4467
0.6	1.072	.9333	1.148	.8710	3.6	1.514	.6607	2.291	.4365
0.7	1.084	.9226	1.175	.8511	3.7	1.531	.6531	2.344	.4266
0.8	1.096	.9120	1.202	.8318	3.8	1.549	.6457	2.399	.4169
0.9	1.109	.9016	1.230	.8128	3.9	1.567	.6383	2.455	.4074
1.0	1.122	.8913	1.259	.7943	4.0	1.585	.6310	2.512	.3981
1.1	1.135	.8810	1.288	.7762	4.1	1.603	.6237	2.570	.3890
1.2	1.148	.8710	1.318	.7586	4.2	1.622	.6166	2.630	.3802
1.3	1.161	.8610	1.349	.7413	4.3	1.641	.6095	2.692	.3715
1.4	1.175	.8511	1.380	.7244	4.4	1.660	.6026	2.754	.3631
1.5	1.189	.8414	1.413	.7079	4.5	1.679	.5957	2.818	.3548
1.6	1.202	.8318	1.445	.6918	4.6	1.698	.5888	2.884	.3467
1.7	1.216	.8222	1.479	.6761	4.7	1.718	.5821	2.951	.3388
1.8	1.230	.8128	1.514	.6607	4.8	1.738	.5754	3.020	.3311
1.9	1.245	.8035	1.549	.6457	4.9	1.758	.5689	3.090	.3236
2.0	1.259	.7943	1.585	.6310	5.0	1.778	.5623	3.162	.3162
2.1	1.274	.7852	1.622	.6166	5.1	1.799	.5559	3.236	.3090
2.2	1.288	.7762	1.660	.6026	5.2	1.820	.5495	3.311	.3020
2.3	1.303	.7674	1.698	.5888	5.3	1.841	.5433	3.388	.2951
2.4	1.318	.7586	1.738	.5754	5.4	1.862	.5370	3.467	.2884
2.5	1.334	.7499	1.778	.5623	5.5	1.884	.5309	3.548	.2818
2.6	1.349	.7413	1.820	.5495	5.6	1.905	.5248	3.631	.2754
2.7	1.365	.7328	1.862	.5370	5.7	1.928	.5188	3.715	.2692
2.8	1.380	.7244	1.905	.5248	5.8	1.950	.5129	3.802	.2630
2.9	1.396	.7161	1.950	.5129	5.9	1.972	.5070	3.890	.2570

Decibel Table (Cont.)

dB	Current or voltage ratio Gain	Current or voltage ratio Loss	Power ratio Gain	Power ratio Loss	dB	Current or voltage ratio Gain	Current or voltage ratio Loss	Power ratio Gain	Power ratio Loss
6.0	1.995	.5012	3.981	.2512	10.0	3.162	.3162	10.000	.1000
6.1	2.018	.4955	4.074	.2455	10.1	3.199	.3126	10.23	.09772
6.2	2.042	.4898	4.169	.2399	10.2	3.236	.3090	10.47	.09550
6.3	2.065	.4842	4.266	.2344	10.3	3.273	.3055	10.72	.09333
6.4	2.089	.4786	4.365	.2291	10.4	3.311	.3020	10.96	.09120
6.5	2.113	.4732	4.467	.2239	10.5	3.350	.2985	11.22	.08913
6.6	2.138	.4677	4.571	.2188	10.6	3.388	.2951	11.48	.08710
6.7	2.163	.4624	4.677	.2138	10.7	3.428	.2917	11.75	.08511
6.8	2.188	.4571	4.786	.2089	10.8	3.467	.2884	12.02	.08318
6.9	2.213	.4519	4.898	.2042	10.9	3.508	.2851	12.30	.08128
7.0	2.239	.4467	5.012	.1995	11.0	3.548	.2818	12.59	.07943
7.1	2.265	.4416	5.129	.1950	11.1	3.589	.2786	12.88	.07762
7.2	2.291	.4365	5.248	.1905	11.2	3.631	.2754	13.18	.07586
7.3	2.317	.4315	5.370	.1862	11.3	3.673	.2723	13.49	.07413
7.4	2.344	.4266	5.495	.1820	11.4	3.715	.2692	13.80	.07244
7.5	2.371	.4217	5.623	.1778	11.5	3.758	.2661	14.13	.07079
7.6	2.399	.4169	5.754	.1738	11.6	3.802	.2630	14.45	.06918
7.7	2.427	.4121	5.888	.1698	11.7	3.846	.2600	14.79	.06761
7.8	2.455	.4074	6.026	.1660	11.8	3.890	.2570	15.14	.06607
7.9	2.483	.4027	6.166	.1622	11.9	3.936	.2541	15.49	.06457
8.0	2.512	.3981	6.310	.1585	12.0	3.981	.2512	15.85	.06310
8.1	2.541	.3936	6.457	.1549	12.1	4.027	.2483	16.22	.06166
8.2	2.570	.3890	6.607	.1514	12.2	4.074	.2455	16.60	.06026
8.3	2.600	.3846	6.761	.1479	12.3	4.121	.2427	16.98	.05888
8.4	2.630	.3802	6.918	.1445	12.4	4.169	.2399	17.38	.05754
8.5	2.661	.3758	7.079	.1413	12.5	4.217	.2371	17.78	.05623
8.6	2.692	.3715	7.244	.1380	12.6	4.266	.2344	18.20	.05495
8.7	2.723	.3673	7.413	.1349	12.7	4.315	.2317	18.62	.05370
8.8	2.754	.3631	7.586	.1318	12.8	4.365	.2291	19.05	.05248
8.9	2.786	.3589	7.762	.1288	12.9	4.416	.2265	19.50	.05129
9.0	2.818	.3548	7.943	.1259	13.0	4.467	.2239	19.95	.05012
9.1	2.851	.3508	8.128	.1230	13.1	4.519	.2213	20.42	.04898
9.2	2.884	.3467	8.318	.1202	13.2	4.571	.2188	20.89	.04786
9.3	2.917	.3428	8.511	.1175	13.3	4.624	.2163	21.38	.04677
9.4	2.951	.3388	8.710	.1148	13.4	4.677	.2138	21.88	.04571
9.5	2.985	.3350	8.913	.1122	13.5	4.732	.2113	22.39	.04467
9.6	3.020	.3311	9.120	.1096	13.6	4.786	.2089	22.91	.04365
9.7	3.055	.3273	9.333	.1072	13.7	4.842	.2065	23.44	.04266
9.8	3.090	.3236	9.550	.1047	13.8	4.898	.2042	23.99	.04169
9.9	3.126	.3199	9.772	.1023	13.9	4.955	.2018	24.55	.04074

Decibel Table (Cont.)

dB	Current or voltage ratio		Power ratio		dB	Current or voltage ratio		Power ratio	
	Gain	Loss	Gain	Loss		Gain	Loss	Gain	Loss
14.0	5.012	.1995	25.12	.03981	17.5	7.499	.1334	56.23	.01778
14.1	5.070	.1972	25.70	.03890	17.6	7.586	.1318	57.54	.01738
14.2	5.129	.1950	26.30	.03802	17.7	7.674	.1303	58.88	.01698
14.3	5.188	.1928	26.92	.03715	17.8	7.762	.1288	60.26	.01660
14.4	5.248	.1905	27.54	.03631	17.9	7.852	.1274	61.66	.01622
14.5	5.309	.1884	28.18	.03548					
14.6	5.370	.1862	28.84	.03467	18.0	7.943	.1259	63.10	.01585
14.7	5.433	.1841	29.51	.03388	18.1	8.035	.1245	64.57	.01549
14.8	5.495	.1820	30.20	.03311	18.2	8.128	.1230	66.07	.01514
14.9	5.559	.1799	30.90	.03236	18.3	8.222	.1216	67.61	.01479
					18.4	8.318	.1202	69.18	.01445
15.0	5.623	.1778	31.62	.03162	18.5	8.414	.1189	70.79	.01413
15.1	5.689	.1758	32.36	.03090	18.6	8.511	.1175	72.44	.01380
15.2	5.754	.1738	33.11	.03020	18.7	8.610	.1161	74.13	.01349
15.3	5.821	.1718	33.88	.02951	18.8	8.710	.1148	75.86	.01318
15.4	5.888	.1698	34.67	.02884	18.9	8.811	.1135	77.62	.01288
15.5	5.957	.1679	35.48	.02818					
15.6	6.026	.1660	36.31	.02754	19.0	8.913	.1122	79.43	.01259
15.7	6.095	.1641	37.15	.02692	19.1	9.016	.1109	81.28	.01230
15.8	6.166	.1622	38.02	.02630	19.2	9.120	.1096	83.18	.01202
15.9	6.237	.1603	38.90	.02570	19.3	9.226	.1084	85.11	.01175
					19.4	9.333	.1072	87.10	.01148
16.0	6.310	.1585	39.81	.02512	19.5	9.441	.1059	89.13	.01122
16.1	6.383	.1567	40.74	.02455	19.6	9.550	.1047	91.20	.01096
16.2	6.457	.1549	41.69	.02399	19.7	9.661	.1035	93.33	.01072
16.3	6.531	.1531	42.66	.02344	19.8	9.772	.1023	95.50	.01047
16.4	6.607	.1514	43.65	.02291	19.9	9.886	.1012	97.72	.01023
16.5	6.683	.1496	44.67	.02239					
16.6	6.761	.1479	45.71	.02188	20.0	10.00	.1000	100.00	.01000
16.7	6.839	.1462	46.77	.02138	30.0	31.62	.0316	10^3	10^{-3}
16.8	6.918	.1445	47.86	.02089	40.0	100.00	.0100	10^4	10^{-4}
16.9	6.998	.1429	48.98	.02042	50.0	316.2	.0032	10^5	10^{-5}
					60.0	10^3	10^{-3}	10^6	10^{-6}
17.0	7.079	.1413	50.12	.01995	80.0	10^4	10^{-4}	10^8	10^{-8}
17.1	7.161	.1396	51.29	.01950	100.0	10^5	10^{-5}	10^{10}	10^{-10}
17.2	7.244	.1380	52.48	.01905	120.0	10^6	10^{-6}	10^{12}	10^{-12}
17.3	7.328	.1365	53.70	.01862	140.0	10^7	10^{-7}	10^{14}	10^{-14}
17.4	7.413	.1349	54.95	.01820	180.0	10^9	10^{-9}	10^{18}	10^{-18}

Index

435